Volume 8

REGIONAL GEOGRAPHY

REGIONAL GEOGRAPHY

Current developments and future prospects

Edited by
R.J. JOHNSTON, J. HAUER, AND G.A. HOEKVELD

Routledge
Taylor & Francis Group

LONDON AND NEW YORK

First published in 1990

This edition first published in 2014
by Routledge
2 Park Square, Milton Park, Abingdon, Oxfordshire OX14 4RN

and by Routledge
711 Third Avenue, New York, NY 10017

First issued in paperback 2015

Routledge is an imprint of the Taylor & Francis Group, an informa business

British Library Cataloguing in Publication Data
A catalogue record for this book is available from the British Library

ISBN: 978-0-415-83447-6 (Set)
ISBN 13: 978-1-138-99716-5 (pbk)
ISBN 13: 978-0-415-73485-1 (hbk) (Volume 8)

Publisher's Note
The publisher has gone to great lengths to ensure the quality of this reprint but
points out that some imperfections in the original copies may be apparent.

Disclaimer
The publisher has made every effort to trace copyright holders and would
welcome correspondence from those they have been unable to trace.

Regional Geography

Current developments and
future prospects

Edited by
R.J. Johnston, J. Hauer, and
G.A. Hoekveld

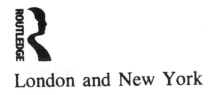

London and New York

First published 1990
by Routledge
11 New Fetter Lane, London EC4P 4EE

Simultaneously published in the USA and Canada
by Routledge
a division of Routledge, Chapman and Hall, Inc.
29 West 35th Street, New York, NY 10001

Printed and bound in Great Britain by Mackays of Chatham PLC, Kent

British Library Cataloguing in Publication Data

Regional geography: current developments and future
   prospects. – (Routledge series on geography and
   environment)
   1. Regional geography
   I. Johnston, R.J. (Ronald John), *1941–*
   II. Hauer, J. III. Hoekveld, G.A.
   910'.091
   ISBN 0-415-05247-5

Library of Congress Cataloging in Publication Data

Regional geography : current developments and future prospects /
   edited by R. Johnston, J. Hauer, and G.A. Hoekveld.
       p. cm. – (Routledge series on geography and environment)
       Papers from a seminar convened by members of the Dept. of
   Geography of the University of Utrecht in October 1987.
   Includes bibliographical references.
   ISBN 0-415-05247-5
   1. Geography–Congresses. I. Johnston, R.J. (Ronald John)
   II. Hauer, J. III. Hoekveld, G.A. IV. Series.
   G56.R42   1990
   910–dc20   89-70149
                                                                  CIP

Contents

Contents

Figures

Figures

Tables

x

Preface

There has been considerable interest shown in recent years in not only reviving but also restructuring the field of regional geography. To explore the nature of such a revival and reconstruction, a seminar was held at Utrecht in October 1987, which was convened by members of the Department of Geography at the University there.

The seminar focused on two main tasks. The first was to review the present state of regional geography in the countries represented at the seminar; the proceedings of that part have been published in L.J. Paul (ed.) (1988) *Post-War Development of Regional Geography*, Netherlands Geographical Studies 86, Amsterdam and Utrecht.

The second task involved exploring the future of regional geography, and was addressed by the majority of the papers. A selection of them have since been revised and are presented here, as a contribution to the continuing debate.

The chapters in the book are of four main types and are organized accordingly. The first three are best described as *critiques*, presenting arguments against what has been done in geography in the past and arguments in favour of developments that can be identified as a 'new regional geography'. They are followed by a further group of three, which are analyses of *possible bases for regional geography*. These deal in turn with world-systems analysis, diffusion studies, and structuration theory. The next two investigate the *rationale for regional geography* — why should its case be advanced? And the final two chapters provide *examples* of the sort of work that might characterize a 'new regional geography'.

In producing this volume, we are grateful to all of those who took part in the organization and conduct of the seminar, and to Joan Dunn for preparing the entire manuscript for production.

R.J. Johnston, Joost Hauer, Gerard A. Hoekveld

1 Region, place and locale: an introduction to different conceptions of regional geography

R.J. Johnston, Joost Hauer, and Gerard A. Hoekveld

In his influential book, *Ideology, Science and Human Geography* (1978), Derek Gregory concluded by arguing that

> Ever since regional geography was declared to be dead —
> most fervently by those who had never been much good at
> it anyway — geographers, to their credit, have kept trying
> to revivify it in one form or another. ... This is a vital task.
> ... We need to know about the constitution of *regional*
> social formations, of *regional* articulations and *regional*
> transformations.
>
> (Gregory 1978:171)

His advocacy had little immediate impact, and the critique of positivism that was presented in the book made a much greater impression than the call for a revised regional geography. In the ensuing decade, however, Gregory's case has been taken up by a number of other authors who have argued for the need not to revive regional geography as it was traditionally conceived but rather to develop a new regional geography that is sensitive to the nuances of areal differentiation and can show how they were central to the operations of most aspects of past and contemporary societies.

The relative decline of regional geography began in the English-speaking world in the 1950s, with the expansion of interest first in topical specialisms and then in the methodological developments that were widely represented as the 'quantitative and theoretical revolution'. Although some adherents of the latter trend argued that they were merely improving the practice of regional geography and its emphasis on synthesis (Berry 1964), in effect regional geography was substantially downplayed (Johnston 1983) and replaced by a focus on regularities in spatial organization and behaviour. That trend was slowly accepted in other countries, where the discipline similarly changed and, among other things, saw the links between physical and human geography become weaker.

Why, then, did Gregory call for a revival of regional geography, little more than two decades after others were seeking to promote its demise? Was it because he wished to bring back the concepts of synthesis and areal differentiation to the top of the geographer's agenda — to be the 'highest form of the geographer's art' (Hart 1982) — because synthesis is the discipline's *raison d'être* in an increasingly fragmented academia? Or was it because the search for universal tendencies by spatial analysts was proving unfruitful, and hence leading geographers into a 'spacious cul-de-sac' (Blaikie 1978)? There can be little doubt that it was the latter much more than the former. Gregory's terminology — with references to 'social formations' and 'transformations', for example — is very different from that of traditional regional geography. Although his Ph.D. thesis (Gregory 1982) has been interpreted by some as little more than a conventional regional historical geography, set within a brief introductory and concluding framework that is somewhat divorced from the empirical material, his goal was certainly innovative. As he expressed it:

> This book is intended as a break with ... traditions. In it, I draw upon social theory to explicate the transformation of the woollen industry of the West Riding of Yorkshire between c. 1780 and c. 1840, and in particular to show how the change from a domestic to a factory system of production ... involved a local transition in human experience and social structure which was tied into much wider congeries of changes in economy, politics and ideology.
>
> (Gregory 1982:2)

Similarly, there can be no doubt that his earlier book was intended and accepted as such by most readers, to be a devastating critique of the practice of geography as it had developed through the 'quantitative and theoretical revolution'. Indeed, he opens it with the statement that:

> In this book I have tried to develop an alternative conception of science on which our inquiries might be based.
>
> (Gregory 1978:11)

What, then, are the criticisms of the type of geography (specifically, human geography) now widely and often derogatively referred to as 'positivist' which lead to the call for a revived, yet new regional geography? And what form should that new regional geography take? These are the central questions addressed by the contributors to this book; the present introduction highlights the issues that are to receive attention.

The need for regional study

Gregory's general argument has been taken up by a number of authors, who have developed the thesis — in a variety of ways — and have provided empirical backing for the argument. Notable among these has been Doreen Massey, whose empirical research for more than a decade focused on under-standing the changing industrial geography of the United Kingdom. She argued that 'geography matters' (Massey 1984a:53), but that neither space nor distance is a crucial independent variable affecting human behaviour. Rather, her case was that the social relations on which the geography of production is based vary spatially; the labour process under capitalism is not the same everywhere, but differs between 'regimes' in ways that reflect their particular histories and influence their futures. As she and others have clearly demonstrated (e.g. Harvey 1982; Smith 1984), uneven development is a necessary characteristic of the capitalist mode of production, but the pattern of uneven development is not fixed. It is rewritten as part of the restructuring that is also a necessary element of capitalism, and each new chapter in the rewriting (what she terms 'layers of development') reflects the interaction of general processes with specific regional contingencies (Massey 1984b).

How do we approach the study of those regional contingencies? And why are they there in the first place? With regard to the latter, it is argued that the capitalist mode of production has been superimposed upon, and has thus incorporated, a great variety of pre-capitalist modes, where the local *modus vivendi* was a function of the local physical environment, as the residents had come to appreciate it. It was modified through centuries of accommodating to environmental change and contacts with other regional groups (forced or otherwise) who had developed slightly different ways of organizing for survival. Thus the pre-capitalist map of the world was a very rich mosaic of regional cultures, of local organizations created to promote collective survival, passed on between generations, and slowly modified in response to contemporary events.

3

As the capitalist mode of production spread out from its origins in north-western Europe, so it incorporated an increasing variety of such local cultures. In part they were subjugated, because their structures were counter-productive to the capitalist goal, but many elements were retained, since capitalism is a very robust mode of production that can accept a wide variety of practices as long as they do not prevent (nor substantially retard) the goal of wealth accumulation. Thus we find the great range of local cultures today, all embracing capitalism and advancing its general goals.

The need for regional geography, then, is a need to understand the contexts within which the capitalist mode of production has expanded and is practised today. (The same case can be made for socialism, too, as Kuklinski's [1987] work on Poland shows.) To appreciate how capitalism operates today, we need to understand the milieux within which it was nurtured, and the milieux that it has created. To understand our present and our future, we must understand their origins in a spatially differentiated past. And, most importantly of all, since uneven development (= areal differentiation = geography) is apparently a necessary component of the capitalist mode of production, we must appreciate how that is created.

What method to use?

How are we to do that? Is it through a series of case studies, out of which general understanding emerges in an inductivist way? Is it by the development of some grand theory that can tell us how geography is created and recreated?

The last two decades have seen geographers actively searching for a theoretical framework that will allow them to develop a general understanding of the processes of uneven development, with which they can situate their empirical materials. Within their discipline, the most attractive such framework has been that developed by Hägerstrand (1968) under the rubric of *time-geography*. Initially, this was interpreted as a 'physicalist' model for the study of behaviour constrained by space and time (Thrift 1977); the emphasis was on tracing people's paths through time and space and appreciating their personal development by understanding the contexts in which they resided (as shown in Pred's [1979; 1984] autobiographical writings). But Hägerstrand and his sympathizers showed that the project was much more subtle and wide-ranging, concerned with understanding situations (that could be interpreted as local cultures) and their meanings to those present.

Outside geography, a theoretical framework that has attracted considerable attention is that of *structuration*, developed single-handedly by the Cambridge sociologist Anthony Giddens (see especially Giddens 1984). Giddens's goal is to break down the dichotomy, which is typical of much social theory, between the structure of society and the behaviour of individual agents. Neither is independent of the other: the structures of society are created by human actions; humans are then socialized in the context of those structures; their actions reflect the structures; in acting, they recreate the structures. Thus there is a continuous reflexive interrelationship between structure and agent: the structure both constrains and enables action (it provides the resources with which agents act, and without which they could not respond knowingly to situations, but the range of resources that is provided constrains them to certain actions and precludes others); the actions recreate the structure, perhaps not in exactly the same form.

The structures to which Giddens refers are social systems that bind people together in order to enable them to live routine, everyday lives. Those systems are sets of rules that are evolved so that everyday conduct is routinized, and people are integrated into the local system of rules. Such routinized integration requires people to be co-present in time and space, hence the organizing units of societies are what Giddens terms *locales*, or regionalized social systems. Thus the processes of structuration, the recursive interrelationships between structures and agents, take place in particular spatial contexts (which we might term *regions*). Rules apply in *places*.

In complex societies, the rules are many and varied, and major elements of the social system comprise institutions that are created to make and enforce rules: these institutions form the state apparatus (Clark and Dear 1984). To some theorists of the state, the imposition of those rules requires a particular type of locale — a bounded territory within which the state apparatus is sovereign (Mann 1984). According to their arguments, *territoriality* as a strategy for imposing rules is not only efficient, as Sack (1986) would argue, but also necessary. Hence, if states are necessary to capitalism, then so are particular forms of locale.

Giddens's theory has much in common with Hägerstrand's, as he realizes (Giddens 1985). Others have enthusiastically built upon this commonality (Pred 1985; 1986). This approach is not without its critics (e.g. Gregory 1985; Gregson 1986; 1987a), yet they all accept the general arguments regarding the spatial specificity of social systems and the spatio-temporal constraints to human agency that both Giddens and Hägerstrand identify.

One issue that several commentators have raised concerns the conduct of research within a structuration context; where does one break into a continually reflexive process (Gregson 1986)? Gregory (1980) has argued that agency must come first, that all structures are social creations. Hence, full understanding requires historical depth, (as argued in Johnston 1989). Where do we get that depth?

To Harvey (1987), that depth comes from what he identifies as the only valid theoretical framework, Marx's *historical materialism*. This provides the means for understanding the origins of the capitalist mode of production, the crises inherent to that mode, and the necessity of both uneven development and the spatial restructuring of that pattern of uneven development (Harvey 1982). But his approach, despite the claims of some critics of structural Marxism (Duncan and Ley 1982), is far from deterministic. He accepts that the details of locales will influence the processes of restructuring that produce and reproduce uneven development and he shows, through his studies of Baltimore and Paris (Harvey 1985), how detailed appreciation of local conditions influences the realization of general trends.

Harvey's work connects to, but is also clearly distanced from, two other theoretical strands. The first is *realism*, a philosophy for social science that identifies three separate domains of study (Sayer 1984): the real, which are the structures containing the causal mechanisms (such as the need to accumulate under capitalism); the empirical, which are the outcomes of the operation of those mechanisms; and the actual, which are the events that result from interpretations of the real and that produce the outcomes, such as patterns of industrial investment. (For an expansion, see Johnston 1986). Thus one has general causal processes, but these are operated in contingent situations, the local social systems which comprise local cultural interpretations of how those processes should be enacted. Finally, there is Wallerstein's *world-systems project* (Taylor 1985a; 1985b), which seeks to understand global capitalism within a Marxist framework, and into which Taylor (1982) has incorporated an explicit spatial referencing.

In both realism and world-systems analysis there is clear recognition that the causal mechanisms that are at the heart of the structures being considered are: (a) human creations; (b) operated through individual decision-making; and therefore (c) influenced in their outcomes by local contingent circumstances. People interpret the mechanisms according to their views of how the structure should be worked and these views are created by their socialization. This does not mean that their views, once developed, are fixed. It does mean three things. First they learn about the structures — often implicitly rather than explicitly — as part of their education, broadly defined. Second, in their activities, both routine and non-routine, they put the results of their learning into action, in some cases after considering what should be done in a particular context, and therefore they extend their interpretation of the structures. Such consideration may involve their taking the advice of others, either directly or indirectly (i.e. through books). Third, the consequences of their activities become part of the context within which others are socialized in the future. Thus agents are at the centre of a web of influences; they have available to them as resources both the set of ideas that comprises their socialized understanding of the structures that they are operating and their perceptions of the empirical reality within which they are to act.

If the web of influences varies spatially, then we have the basis for a spatially variable set of realizations of the underlying structure — whether capitalist, socialist, or whatever. Why should the web be spatially variable? First, there is the straightforward geographical fact that places differ physically. Although there is not necessarily a one-to-one relationship between the actual physical differences and how they are perceived, nevertheless the nature of those physical differences means that people are constrained to interpret the structural imperatives (especially those relating to surviving by winning food, water and shelter from the environment) differently in different places. Second, out of those differences have arisen different cultures, different sets of collective ideas (into which people are socialized), about how to survive and to organize social life. These cultures, as outlined earlier, form the basic geography on to which the capitalist mode of production has been superimposed, and are the basics of the different local interpretations of how capitalism should be operated. Those interpretations, as they are enacted, produce a well-developed local context, a set of contingent circumstances, within which capitalism operates. Not all aspects of the culture are directly related to the basic imperatives of the mode of production, of course, because they cover a wide range of social and political activity outside the sphere of the economic. But few of those aspects are entirely dissociated from the economic — as the works of Doreen Massey and others show.

7

All of these theoretical frameworks recognize the importance of place, therefore, and accept that geographical variability in the operation of modes of production is extremely likely, if not necessary. People create structures in the context of places; those structures then condition the making of people. In that recursive process, people and places change, continually. The places can change in terms of both their content and their extent, because distance, while it may be an impediment to the spread of ideas in certain contexts, is rarely an impervious barrier. Thus the people-place interaction incorporates a place-place (or people-in-place/people-in-place) interaction. Places and people change according to external as well as internal linkages.

What we lack in these theoretical frameworks is a clear set of methodological protocols. How do we conduct research in their contexts? Sayer (1984) and others have suggested a methodology for realism in general terms, and others have sought to apply it (e.g. Allen 1983). The empirical studies of Massey (1984b) provide exemplars also, but there is clearly much to be done in extending the methodology, and thereby consolidating the theoretical frameworks. It is towards that end that the essays in this book are directed.

In summary

'Regional geography is dead: long live regional geography' may well be presented as the catch-phrase of the argument that we have developed here. The regional geography of the period prior to 1960 has been rejected by many as an unsatisfactory approach to the discipline. But increasingly geographers have realized that this rejection of a particular approach should not be taken to imply a rejection of the need to study regions. Regions matter; places matter; locales matter. They are the contingent circumstances in which people are made, and in which they act as agents within the structures that are our ways of organizing life for ourselves on earth. So how are those contingent circumstances created, and how do people interact with them?

References

Allen, J. (1983) 'Property relations and landlordism: a realist approach', *Environment and Planning, D: Society and Space* 1:191-203.
Berry, B.J.L. (1964) 'Approaches to Regional Geography: A Synthesis', *Annals of the Association of American Geographers* 54:2-11.
Blaikie, P.M. (1978) 'The Theory of the Spatial Diffusion of Innovations: A

Spacious Cul-de-sac', *Progress in Human Geography* 2:268-95.

Clark, G.L. and Dear, M.J. (1984) *State Apparatus*, Boston: Allen & Unwin.

Duncan, J.S. and Ley, D. (1982) 'Structural Marxism and Human Geography: A Critical Assessment', *Annals of the Association of American Geographers* 72:30-59.

Giddens, A. (1984) *The Constitution of Society*, Cambridge: Polity Press.

— (1985) 'Time, Space and Regionalisation', pp. 265-95 in D. Gregory and J. Urry (eds) *Social Relations and Spatial Structures*, London: Macmillan.

Gregory, D. (1978) *Ideology, Science and Human Geography*, London: Hutchinson.

— (1980) 'The Ideology of Control: Systems Theory and Geography', *Tijdschrift voor Economische en Sociale Geografie* 71:327-42.

— (1982) *Regional Transformation and Industrial Revolution: A Geography of the Yorkshire Woollen Industry*, London: Macmillan.

— (1985) 'Suspended Animation: The Stasis of Diffusion Theory', pp. 296-336 in D. Gregory and J. Urry (eds) *Social Relations and Spatial Structures*, London: Macmillan.

Gregson, N. (1986) 'On Duality and Dualism: The Case of Structuration and Time Geography', *Progress in Human Geography* 10(2):184-205.

— (1987) 'Structuration Theory: Some Thoughts on the Possibilities for Empirical Research', *Environment and Planning, D: Society and Space* 5(1):73-92.

Hägerstrand, T. (1967) *Innovation Diffusion as a Spatial Process*, Chicago: University of Chicago Press.

Hart, J.F. (1982) 'The Highest Form of the Geographer's Art', *Annals of the Association of American Geographers* 72:1-29.

Harvey, D. (1982) *The Limits to Capital*, Oxford: Basil Blackwell.

— (1985) *Consciousness and the Urban Experience*, Oxford: Basil Blackwell.

— (1987) 'Three Myths in Search of a Reality in Urban Studies', *Environment and Planning, D: Society and Space* 5:367-76.

Johnston, R.J. (1983) 'The Region in Twentieth Century British Geography', *History of Geography Newsletter* 4:26-35.

— (1986) *On Human Geography*, Oxford: Basil Blackwell.

— (1987) *Geography and Geographers*, London: Edward Arnold.

— (1989) 'Economic and Social Policy Implementation and Outputs: An Exploration of Two Contrasting Geographies', in J. Kodras and J.P. Jones (eds) *Spatial Dimensions of US Social Policy*, London: Edward Arnold.

Kuklinski, A. (1987) 'Local Studies in Poland: Experiences and Prospects', pp. 7-16 in P. Dutkiewicz and G. Gorzelak (eds) *Local Studies in Poland*, Warsaw: Institute of Space Economy, University of Warsaw.

Mann, M. (1984) 'The Autonomous Power of the State: Its Origins, Mechanisms and Results', *European Journal of Sociology* 25:185-213.

Massey, D. (1984a) 'Introduction: Geography Matters', pp. 1-11 in D. Massey and J. Allen (eds) *Geography Matters!* Cambridge: Cambridge

University Press.
— (1984b) *Spatial Divisions of Labour*, London: Macmillan.
Pred, A.R. (1979) 'The Academic Past through a Time-Geographic Looking Glass', *Annals of the Association of American Geographers* 69:175-80.
— (1984) 'From Here and Now to There and Then: Reflections on Diffusions, Defusions and Disillusions', pp. 86-103 in M. Billinge, D. Gregory, and R. Martin (eds) *Recollections of a Revolution*, London: Macmillan.
— (1985) 'The Social becomes the Spatial and the Spatial becomes the Social', pp. 336-75 in D. Gregory and J. Urry (eds) *Social Relations and Spatial Structures*, London: Macmillan.
— (1986) *Place, Practice and Structure*, Cambridge: Polity Press.
Sack, R.D. (1986) *Human Territoriality*, Cambridge: Cambridge University Press.
Sayer, A. (1984) *Method in Social Science*, London: Hutchinson.
Smith, N. (1984) *Uneven Development*, Oxford: Basil Blackwell.
Taylor, P.J (1982) 'A Materialist Framework for Political Geography', *Transactions, Institute of British Geographers* NS7:15-34.
— (1985) *Political Geography: World-Economy, Nation-State and Locality*, London: Longman.
— (1985) 'The World-Systems Project', pp. 269-88 in R.J. Johnston and P.J. Taylor (eds) *A World in Crisis?: Geographical Perspectives* Oxford: Basil Blackwell.
Thrift, N.J. (1977) *An Introduction to Time-Geography*, CATMOG 13, Norwich: Geo Books.

2 Regional geography must adjust to new realities

Gerard A. Hoekveld

Introduction

> But an investigation which is successful in identifying and presenting regions must seek meaningful patterns and should contain the demonstration that the patterns presented are, in fact, significant. ... The regional pattern has both meaning and significance when it can be interpreted in terms of systematically related processes, operating through time ...
>
> (Whittlesey 1954:33)

Regional geography is back, they say. But why? No more than a quarter of a century ago, it was dismissed by some geographers with fervour, blame, and contempt, while other geographers indifferently let it slip into oblivion. The first group regarded regional geography as an outmoded form of description which could detract from the newly won status of geography as a spatial and social science. The second group was so rapt by the fascinating horizons of geographical research in topical subdisciplines, such as urban, rural and economic geography, that traditional regional geography seemed to disappear from their view. Yet in Germany and France regional monographs continued to be written and published despite the debates. Why, then, is there a call for the revival of regional geography? Have the principal objections against it, raised in the 1960s and 1950s, been met by our generation? They have not. There is still a lack of theory to guide the selection of phenomena and the relationships between them, accompanied by methodological arbitrariness. There is no consensus on what regional monographs should contribute to progress in regional geographical research. There are other disciplinary weaknesses, as well: too much emphasis on the unique and singular; too little generalization and therefore no applicability of new insights to other regions; not

11

enough use of the accepted research methodology and techniques of the other social sciences; an overabundance of eclecticism and unfounded choices of 'problems', 'themes' and 'interdependencies'; and, especially in Germany and France, too much research aimed at the study of landscapes and not enough at the study of relevant social problems and phenomena (see, for instance, Stewig 1979:5). It is remarkable that the methodological groundwork that was done by the Association of American Geographers (AAG) commission, which was reported by D. Whittlesey in 1954 and, for instance, by Grigg in his study about logic and regional systems in 1965, was not followed up, although this work is still relevant.

It seems that pleas for a revival of regional geography often ignore the scientific reasons for its deplorable state. It is argued that geography needs regional geographical studies in order to maintain public relations for geography. Regional geographers should present their studies about places and regions as attractive and informative reading matter for a general public. In particular, regional geography should conquer the press and the 'new media'. These pleas, in fact, call for applied regional geographies. Such applications will derive their form, and partly their content, from the purpose of these applications. It makes a big difference whether one has to depict a region for the general television-viewing public, for a geography class of 14-year olds, for a planning commission, or for a board of directors considering their firm's investments. This means that various types of regional geography are needed but, first, there must be an academic, scientific regional geography. A strong methodological basis should counter the objections raised by too many members of the scientific geographical community. This chapter is confined to the academic regional geography. This brings us to the problem of concepts in and of regional geography.

Concepts and theories in regional geography

One of the major shortcomings of regional geographers is supposed to be their failure to develop and use a conceptual apparatus and theories, in contrast to geographers devoting themselves to topical studies. However, classical French and German regional geographies were certainly based on geographical paradigms which shaped the theoretical structure and composition of the empirical contents of many monographs. Authors, nevertheless, rarely used intermediate-level theories to connect the paradigm with the description of facts and relations. This situation hardly changed when the geographical paradigm of the relationships between nature and individual was succeeded, first by the paradigm of the genetic of functional landscape physiognomists, later by functionalist studies of spatially framed livelihoods, and still later by the prescript that geographers should tackle 'relevant problems' as the backbone and integrating theme for their studies.

Regional geography is about places, which means areas; it is not about objects which have spatial attributes. The latter are treated in topical geographies. Only the circumstance that the subject of a study is an area makes the study regional geography. This leaves unaffected the idea that regions are mental constructs made by geographers. 'It is an intellectual concept, an entity for the purposes of thought, created by the selection of certain features which are relevant to an areal interest, or a problem, and by the disregard of all features which are considered to be irrelevant' (Whittlesey 1954:30, see also Blaut 1962:2). This implies that regional geographical concepts relate to attributes of areal units, as the AAG commission rightly pointed out in 1954 (Whittlesey 1954:39). Concepts refer to classes of objects; in regional geography, concepts refer to classes of areas. These classes are already generalizations based on common attributes (see Hoekveld-Meijer's chapter in this book). Because a specific area is allocated to a certain class, it is comparable to other areas in the same class. Whittlesey argued that regional geography must be advanced by the use of global classes, as, for instance, Köppen and Thornthwaite did. The studied area must be placed in these more general classes.

Theory, however, and not a 'problem' nor 'an areal interest', must direct the formation of classes and the selection of relationships between attributes of areas in the same class, or in different classes. Theories are the mainstay of a regional geography which is progressing along research frontiers and is developing its specific methodology. Even comparative analysis, considered by many geographers to be a necessary step to theory construction, is already based on theory, as Pickvance (1986) has demonstrated. Stewig (1979:23) proposed that regional geographers proceed on two levels: first, to construct a general theoretical framework or model ('*einen allgemeinen Modellhaften Rahmen*' or a '*Gesamttheorie*'); second, to utilize more limited theories. From these intermediate theories, they can derive hypotheses and explanations for questions or themes which are embedded in the empirical resolution of the general theoretical framework. It was previously mentioned that classical concepts were the same as

13

those which helped to formulate the paradigms. These paradigms provided a general theoretical framework in the absence of a more substantive general theoretical framework and specific theories. For example, Buttimer (1971:194) showed that the paradigm of French classical geography was based on some 'core' or 'umbrella type concepts that were constantly redefined in terms of the changing empirical conditions being studied', and that have been 'analytically dissected' in the fourth, fifth and sixth decades of the twentieth century. It is useful to focus on these concepts before treating the question of whether new concepts are an adequate replacement for the concept of classical regional geography used as overarching general theoretical models.

The key concepts of classical French and German regional geography

In Europe, classical regional geographers (e.g. P. Vidal de la Blache, A.J. Herbertson) conceived of regions as human-centred ecological systems. A later generation (O. Schlüter, C.O. Sauer, and, in certain respects A. Demangeon) saw a region primarily as a landscape, the physiognomy of which was brought about by genetic-morphological processes. The chorological tradition was expressed by A. Hettner:

> The goal of the chorological point of view is to know the character of regions and places through comprehension of the existence together of, and interrelations among, the different realms of reality and their varied demonstrations, and to comprehend the earth's surface as a whole in its actual arrangement in continents, larger and smaller regions and places.
>
> (Hartshorne (trans.) 1959:13)

Since the Department of Geography at the University of Utrecht is traditionally oriented toward French geography, we shall pay special attention to the latter. Ann Buttimer (1971) has aptly described the key concepts of classical French human geography and has placed them in the context of the inspiration which its founder, P. Vidal de la Blache, among others, explicitly received from Ritter and Ratzel. The German Romantic, even Kantian, dichotomy of individual and nature was also at the basis of classical French geography, but it was elaborated in a sophisticated set of related concepts by Vidal. 'Nature' was intertwined with the anthropocentric concept 'milieu': the organically integrated physical and biotic infrastructure of human life on earth (Buttimer 1971:3, footnote 6:45). Vidal followed Ratzel in this organistic concept. He perceived large realms of nature on a global scale, which provided the *'milieux de vie'* of different peoples. World population should be studied in the context of these great *'milieux de vie'* — in other words, of how people have adapted the natural resources of these different *'milieux de vie'* in the creation of *'genres de vie'* or lifestyles (Buttimer 1971:45). This *'genre de vie'* concept is central to Vidal's views. He used it to mean a unified functional pattern of living, which encompassed the ethnographical or cultural component, the historical component, and the ecological perspective on human society, which 'echoed the integration of place, livelihood and social organization in a group's daily life' (Buttimer 1971:53). 'Civilization', another and perhaps the least well-developed of the concepts introduced by Vidal, comprises the world view of a group, its habits, values, attitudes and even psychological characteristics, its 'mentality'. 'Civilization' directs the *'genre de vie'* and its adoption of nature in the *'milieu de vie'*. Thus 'civilization' is the main explanation of the choices that different groups make when they share the same environment.

Like Ratzel, Vidal recognized that every place has a relative position in reference to other places. These relative positions are important because they can promote or hinder human contact and the exchange of people, goods, and ideas. Circulation enables a group to overcome limiting natural conditions. According to Vidal, circulation constitutes a key concept in understanding the emerging urban-industrial world of the twentieth century. 'Region' or *'pays'* or *'contrée'*, is the most complex concept that Vidal introduced (1896). It refers to a large number of traits which influence and transform each other. Their combination was a 'linkage expressing the general laws the terrestrial organism is submitted to'. Vidal and Brunhes, his most renowned successor, had a twofold orientation. They wanted to study generalizations and laws as well as the local *'enchaînement'* of the workings of those laws which generate a unique 'personality' of the 'locale', the *'pays'*. According to Vidal, this personality becomes visible in the physical (natural) physiognomy of the land, the *'paysage'*. With the concept *'paysage'*, Brunhes also implied the imprint of 'civilization' and its accompanying *'genre de vie'*. The *'paysage'* thus represents the face of the region. 'Population', the last key concept, is the basis for the anthropocentric perspective, although Vidal and Ratzel both defined geography as the study of places. Like Ratzel before him, Vidal gave the

15

population characteristics of density, distribution, and movement a central place in his conceptual framework. Brunhes stressed the dynamic aspects of regional life by giving his regional studies a topical design. He retained the concepts of 'milieu' and '*genre de vie*'. The latter concept, however, was defined as a combination of patterns of work, incorporating mentalities, traditions, and so on, as far as they were connected with livelihood. Disregarding the different emphases found among leading French geographers, Figure 2.1a represents their common conceptual framework.

After the Second World War, Vidal's key concepts were analytically dissected. The study of nature and its laws was left to physical geography; more and more, nature was reduced to physical conditions or physical setting. The concept '*genre de vie*' withered away after long debates, although it was still studied in the 1940s and 1950s. This was the result of a combination of adaptions to the local environment and the influences, mostly disruptive, of interaction and communication through which smaller regions became integrated in larger ones. Finally, the systems approach furthered the use of concepts such as economic system, division of labour, geographically significant lifestyles (eventually without a direct tie to livelihood), and spatial organization (Buttimer 1971:187). 'Circulation was extended to serve in its empirical realizations as an index to the character and degree of the organization of space by society, as a promulgator of new ideas, techniques, ambitions and thus as a creative force for regional change' (Buttimer 1971:192). 'Spatial organization' (the '*räumliche Organisation*' of the Germans), although rarely if at all defined, was considered to be a new key concept by Kostrowicki (1975) and by Claval (1984:34). In the 1950s and 1960s, it was increasingly employed to explain the spatial patterns of populations, their resources and artefacts, with the help of theories, e.g. central place and other locational theories, land-rent theories, interaction and behaviourial theories. In this chapter, spatial organization is also seen as a key concept. It is, however, defined here as a whole complex of land use, perceptive, evaluative, ownership, and control relationships between elements of society (e.g. firms, households, governmental agencies) and the parts of the earth's surface which together form an area. These relationships are disclosed in patterns of interaction and communication. Although this spatial organization was mostly interpreted economically, cultural and social geographers used it in ways appropriate to their own fields of investigation.

It has gradually become clear to geographers that capital

plays a very important role in the organization of space. Originally, capital was a local asset in the hands of local landowners or wealthy urban merchants. The development of banks changed this situation (Labasse 1955; 1972). Banking has lost part of its regional autonomy and has become involved in metropolis-based international finance (Thrift 1986:28). Extension of credit to investors and private persons, however, has never pervaded regional life as deeply as nowadays.

The concept 'milieu' was used more and more loosely, and finally served to denote surroundings or physical setting. It tended to be replaced by terms such as region, areal differentiation, or, most often, space, ('*espace*' or '*Raum*'). In this process it lost its materialistic meaning as part of the face of the earth — including its natural and artefactual features and functions as a complex of resources — and became an abstract notion, which was to be filled with substantive variables. The vague but important concept 'civilization' also lost much of its comprehensive meaning. Mostly it was used in an explanatory way in relation to '*genre de vie*' and livelihood, or to society-created forms in the landscape. It was ousted by concepts such as spatial, areal, and functional organization. It cropped up again, however, in the 1970s and 1980s in the context of behaviourial and humanistic geographical studies of social groups and in studies of 'lived space' (the French concept is '*espace vécu*'). The concept 'region' was reduced to mean an area with certain attributes which were relevant only for the purpose of research, e.g. economic regions, experiential space, and so on. With the dissolution of the great classical key concepts, the focus of regional studies was lost and the concept of the region was stripped of its former meanings.

The evolution of the concept of the region in France, from Vidal's 'region' to 'an area of production and consumption, innervated by an urban pattern which is its skeleton' (Dumolard 1980:25) and, further, to a complex spatial category, was most closely paralleled in Germany. There, despite a strong inclination toward the presentation of landscapes or physiognomies of regions, regional geography until very recently was often determined by Hettner's scheme of regional geographical presentation (*das Länderkundliche Schema*). That scheme shows how to compose a regional study, in the form of a book, an article, or an atlas. Hettner identified the realms of abiotic phenomena and their forms, the realms of flora and fauna, and the realm of humanity (settlement, traffic, economy, way of life, ethnic variety, religion and political organization) as the elements of regional description. He insisted, however, that regional geography was

Figure 2.1a The conceptual framework of French classical geography

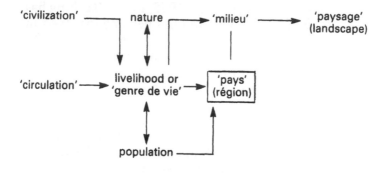

Figure 2.1b The conceptual framework of regional geography in the 1950s and 1960s

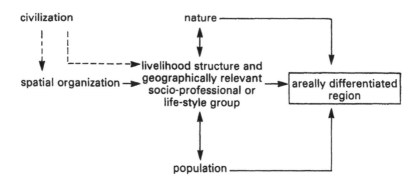

not merely a descriptive activity but also a study of causal relations. After a phase of strong environmentalist inclination, stressing the interconnections between a population and its physical environment, came a long period in which genetically and functionally based landscape morphology was a main topic. In the early 1960s a new phase began which focused on spatial organization (Thomale 1972:203). The beginnings of this functional approach to the study of regions in Germany had already emerged in the 1930s (Bobek, Christaller). After

Figure 2.1c Elements of a new general conceptual framework for regional geography

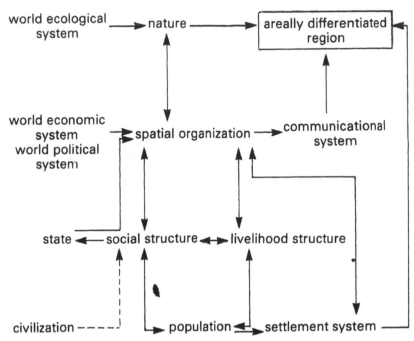

the Second World War the causal, historical, and morphological approaches were gradually replaced by functionalistic approaches, which were soon embedded in systems frameworks (Lichtenberger 1984:163). The same occurred in France.

The recent history of French and German regional geography shows an analytical dissection of the paradigmatic, conceptual framework of the classical phase. This is illustrated in Figures 2.1a–2.1c. Figure 2.1a shows a simplified structure of French classical geographical concepts. The concepts in Figure 2.1b, which in a sense also represents German regional geography in the 1950s and 1960s, are often ambiguous and even more vague than their predecessors. The concepts 'civilization' and landscape are eliminated, while spatial organization is often conceived as an economic concept. 'Civilization' seems to be transformed into concepts such as 'regional identification of actors', 'spatial preferences', and the like. But it is no longer extensively studied on the level of the nation-state and its constituent regions.

Concepts such as 'national character' - still current in the 1930s and 1940s - have been abandoned. In the process, insights were gained into the nature of modern regional systems as urban and industrial systems, or the urban-based organization of life. After a period which studied the spatial organization of regions as interdependent pairs of cities and their hinterlands, there was a progression to the study of regions as interlinked daily urban or rural systems, either as model urban systems or as regional systems with several small nodes (Martin and Nonn 1980).

The results of this bristling development of urban, rural and economic geography, supported by testable theoretical frameworks, were placed at the disposal of regional geography. It was also realized that the vague term 'spatial organization' actually contains many formal organizations which structure the areas of regions in many ways. The same applies to the terms 'population' and 'group', which are comprised of socio-professional and other groups that are used in Figure 2.1. There are many relations between this differentiated population and its highly articulated institutional infrastructure and superstructure. In many cases regions are overlaid by the territorial range of institutions and organizations located elsewhere. Of course, this influences the regional structures and daily systems of the inhabitants. Therefore, organizations should no longer be subsumed under headings such as 'livelihood'. They connect regions and locales with nation-states and sometimes with the world. The institutional system, in particular the many elements of the state, is part of the 'superstructure of society' (Johnston 1982:283).

The analytical dissection of the conceptual framework of the classical paradigms deprived regional geography of its 'grand theory' in France as well as in Germany. Functionalism damaged both the German '*Länderkundliche Schema*', based on claims of cause and effect, and the classical French idea of '*enchaînement*'. The wealth of diverse theories concerning regional dimensions of production, housing, distribution, management and politics, group behaviour, and so on, did not provide a basis for a comprehensive framework. It did offer theoretical touchstones for the study of certain classes of phenomena within a region. In the 1970s and 1980s the theoretical discussion about the fundamental characteristics of scientific knowledge in general and of society in particular sparked debate between 'positivists', 'liberals', 'humanists', 'Marxists', and 'structuralists'. In particular, the discussions of the last two groups, to which R. Hudson, J. Hauer, and R. Lee contribute in this book, are generating new paradigmatic conceptual frameworks for regional geography. The old general theoretical framework, however, did not succumb to the rigidity of formal conceptual analysis, nor to this contemporaneous debate; changing realities made it obsolete.

New regional realities and potential key concepts

The regional realities described by classical French and German geography were those of more or less static rural societies, which were served by urban centres and were primarily based on resources of either their own territories or of horizontally integrated regional industrial complexes.

During the past thirty years, historians, sociologists, and economists have disclosed a vivid picture of the processes which shaped our western dynamic, urban, neo-industrial societies and consequently the changes in spatial organization. Their models of these processes are succinctly summarized by the unfolding of capitalism in the context of emerging world-economies based on urban mercantile and industrial systems (Wallerstein 1974; Braudel 1979; de Vries 1984). Nowadays the structure of world industrial production is no longer organized along the lines of product specialization and regional linkages, but primarily along the lines of transnational vertical integration (Grotewald 1971; Humbert 1986; Dicken 1986). Modern industrialization may be understood by studying relevant actors in industry, commerce, finance (multinationals), and government in the contexts of their own institutional and communication networks. Side by side with this new organization of production, a new hierarchy of world-cities is asserting itself in the existing national city-systems (Friedmann 1986; Brunn and Williams 1983). Twentieth century regional realities are based on quite another type of environment from that which the traditional conceptual frameworks assumed.

Relationships between site, national resource bases, and society are mediated by an international economy, a nation-state, world-cities, and national city or settlement systems, in addition to wide institutional and communication networks and financing organizations such as banks, pension funds, etc.

On the bases of these new regional realities, regional geographers might try to hew new building blocks and assemble a new general conceptual framework to replace the redundant ones. Regions are indeed theoretical constructs. However, these constructs have to be tested in reality. In the scheme of Figure 2.1c, some elements of a new general conceptual framework are suggested. This scheme corresponds to a large extent with Johnston's idea (1986:112-13) that a framework for studying the world as a 'mosaic of places' is provided by the context of the regions. The elements of this framework are summarized below.

World-system

World-system is the first key concept in the framework which might be productive. It is another term for 'the context of regions' and consists of interrelated economic, political, social, and cultural activities. Or, more narrowly defined, an economic world-system is conceived as the partially interconnected networks of market relationships, concerning raw materials, products, labour, and capital, that are maintained by commercial or state-owned enterprises which relate parts of the globe to an international division of labour (see the chapter by Terlouw in this book). On ideological grounds, some authors equate the economic world-system, based on capitalism, with the political world-system. It would be wiser, however, to treat the latter as an independent system. Its actors may pursue aims in the economic world-system that differ from their aspirations although both groups of actors probably have many other interests in common.

Spatial organization

Spatial organization remains an important key concept. Notwithstanding its holistic implications, it is suited to analytical and empirical application. It also includes the institutional superstructure as far as rules and procedures are concerned. Spatial organization is primarily a relational concept. It refers to relations which manifest themselves in spatial patterns, such as land use, daily urban system, and jurisdictional patterns. The second of these patterns is the modern version of a part of the old 'circulation' concept.

Population

Population is the third key concept which must be included in the conceptual framework. It is true that the organizational elements which make up spatial organization are manifestations of the population of an area. Nevertheless, 'population;, conceived as an aggregate of individuals, should be distinguished from the organizational modes in which populations participate and express themselves. Its areal distribution, density, and movements are quite different from the spatial organization of an area. Moreover, regional geographers are students of areal traits, primarily, and not of a population's geographical characteristics in their own right.

Social structure

Social structure, the fourth key concept, is inserted in this scheme because the definition of spatial organization that is given here assumes that the society has a social structure composed of socially related elements which maintain relationships with a territory. These elements can be categorized or grouped in many ways, e.g. in terms of social class, and so on. Spatial organization encompasses the relationships themselves. If social structure is defined broadly, it comprises the traditional livelihood structure (in terms of the system of production) or the employment structure (from the population point of view). But if social structure is defined more narrowly — and this seems most appropriate — then the livelihood structure or productive system must be a separate concept. If a whole country is the subject of a regional geographical study, then the state is an element of the social structure. If, on the other hand, part of a county (i.e. a region) is the subject of study, then the state is a separate key concept which should be included in the scheme. In that case it has the same contextual status as a world-system and often serves as an intermediary between world-systems and regional spatial organization.

Settlement system

Settlement system is the fifth key concept that regional geographers should use. It comprises not only cities but also the smallest hamlets and even isolated dwellings. It is the localized base of populations and the material shelter of industries. It includes important aspects of a society, be they financial, institutional, or even emotional. It needs an institutional network for its continued existence. It forms the spatial link between the level of analysis of the spatial organization, the levels of aggregated phenomena which may form an abstract, regional structure, and the levels of analysis of the actors and their activities. People live — which means that they work, recreate, sleep, and so on — within their settlements. They extend their activity spaces and locate their daily systems between and within settlements.

Figure 2.2a The percentages of the addresses per municipality in The Netherlands that received a daily newspaper in 1986. Values of more than 100 per cent indicates that in some or many households more than one newspaper is delivered

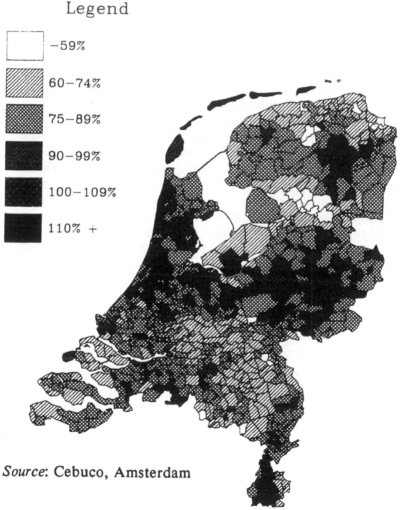

Legend

- −59%
- 60−74%
- 75−89%
- 90−99%
- 100−109%
- 110% +

Source: Cebuco, Amsterdam

Communication system

Communication system, the sixth key concept, should be distinguished from the settlement system, although they share parts of their material infrastructure, such as highways, railroads, and canals. Information channels, such as language- or region-based television networks, regional and national newspapers (see Figure 2.2), as well as religious organizations,

Figure 2.2b The share of the national daily newspapers in the
total of the daily delivered newspapers per municipality in The
Netherlands in 1986

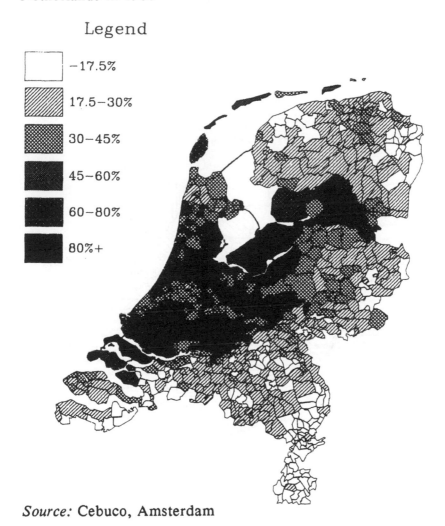

Legend

☐	−17.5%
▨	17.5−30%
▨	30−45%
■	45−60%
■	60−80%
■	80%+

Source: Cebuco, Amsterdam

have an impact on the areal differentiation of the region via
their influence on the perception, attitudes, and values of the
population groups and their actors.

The communication system is a necessary condition for
'regional identity'. The areal differentiation of The
Netherlands (see Figure 2.2) illustrates this statement. There
is a public consciousness of differences between regions in
The Netherlands, although it is a unitary state with a high
degree of regional and local integration in a common national

Figure 2.2c The municipalities in which a specific regional daily newspaper has a share of 25 per cent or more of the total of the delivered daily newspapers in 1986

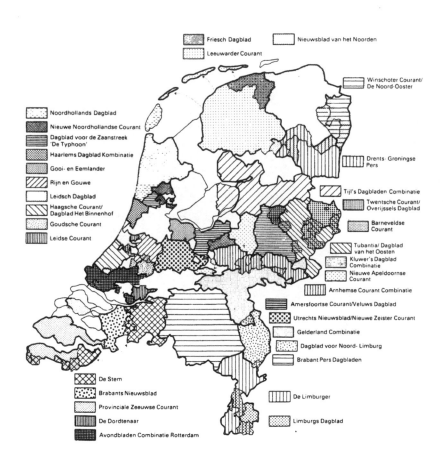

Source: Cebuco, Amsterdam

culture. Before the Second World War the differences were at least partially ascribed to the regional character of their populations. This psychological approach has proved to be untenable. Perhaps the study of the communication system and of certain categories of information might help us to understand regional identity. The information about a particular region and the rest of the country is mediated by national TV as well as by national and regional newspapers, each with its own emphasis and selection of news. In The Netherlands, some areas, particularly in the urbanized and suburbanized western part of the country and in relatively isolated places, are served by national newspapers. Other areas are oriented towards regional newspapers. Regional newspapers choose to inform their readers about their own region more than about other parts of the country. This is both the result and the source of regionalism.

Nature

Nature is the seventh key concept which, perhaps to the surprise of many, should be part of the regional geographical conceptual framework. 'Nature' in a region is part of the world's ecological system as is increasingly becoming clear (again) as recently has been argued by the Club of Rome (1972) or in the Brundtland Report (World Commission on Environment and Development 1987). In regional geography nowadays it is still conceived in a more limited way. The links between livelihood and nature have been endlessly stretched and are obscured by many mediators, such as economic, technological, institutional, and evaluative factors. It is perceived merely as a price-tag on building activities. Even if the relations between nature and spatial organization are no longer felt, the relations with the settlement system, including its infrastructure, are highly visible. At the same time, nature may be a factor in the differentiation of areas, either in its own right, or mediated by the spatial organization.

Civilization

Civilization is the last and most problematic key concept. Its place among the other key concepts is disputable. In regional geography, culture has received a very uneven treatment. Sometimes culture was seen as nothing more than a modest factor explaining some distinguishing characteristics between studied societies or communities. In the years before the Second World War, many studies were written about the 'national character' or the 'character of a community'. The specialization of geography during the 1950s and 1960s was undoubtedly indispensable to the admission of geography to the ranks of sciences with a respectable methodology and status. However, it has deprived geography of the traditional ways of describing civilizations as attributes of regions or localities. National character was not well suited to geographical analysis and its study was suspected of prejudice or caricature. Anti-semitism and nationalism demonstrated the dangers of the unsophisticated use of these images during the Third Reich. The study of national character was left to cultural anthropologists and social psychologists. Perhaps the ideological split into liberal, socialist, and Christian-democratic stances, in societies where cultural differentiation had been reduced to uniformity by nation-states, diverted the interest of researchers to the socio-political attitudes and behaviour of groups, defined in terms of profession and life style. It is interesting that, simultaneously, the cultural perspective in geography narrowed as a consequence of 'spatialization'. For instance, the community studies tradition in the Anglo-American world succumbed to the attacks of the 'sociologicalization' movement in the social sciences in the 1950s and 1960s. Possibly with the help of the study of social structure, institutions, and communications, the concept of civilization might again be received into regional geography (Buttimer and Claval 1987:222).

The conceptual framework sketched above should be seen as a model of the historical development context of regional emergence and regional transformation or development. The areal differentiation which geographers observe depends on their selection of areal attributes. If the observer takes an anthropocentric stance, the areal differentiation will demonstrate the significance of cause and effect in mutual relationships and their impact on the triad of social structure -livelihood-spatial organization. This development is locally embedded in nature and in settlement systems and it functions through communication systems. The workings of this triad are influenced by contextual world-systems, state, and civilization, and perhaps by the quantitative developments of populations and by nature. As the outcome of all this, the areal differentiation is changing in the course of historical processes. Consequently, the potential to discern meaningful regions in the complex real world is changing too. Regional geography should concentrate on the theories that explain processes of areal differentiation. These theories must be formulated at the structural level as well as at the level of actors.

Summary and conclusions

A revival of regional geography presupposes the rebuilding of the general conceptual frameworks that were lost when the paradigmatic umbrella concepts of 'classical' regional geography were dissected. There is an incipient consensus about the contours of this renewed subdiscipline of geography, owing to empirically founded insights. Some concepts have been proposed which might serve as building blocks for this much-needed conceptual framework (see Figure 2.1c). However, no attempt was made to define the nature of the links between these concepts. At the moment it is difficult to find consensus about the kinds and directions of these links because of the ideological — or is it philosophical? — disagreement among geographers. (Marxists probably would prefer separate concepts, such as class structure and mode of production, to articulate the holistic concepts of social structure, social organization, and livelihood and would give the mode of production a more central position). The problem of the relationships within the proposed general conceptual framework concerning regions, e.g. the demographic qualities of the population of the region, its housing markets, and its labour markets (which both influence its settlement system) were not even touched upon. The available theories about areal differentiation seem to be insufficient. This means that the existing theories in thematic geography at least need adjustment. These theories formulate relations between classes of located objects and their attributes, of which the region is but one. Yet the regional geographer's main concern is a theory about the relations between areal classes and their attributes which are, among other things, located objects. The drafting of an all-encompassing general theoretical framework of key concepts may guide the selection of thematic geographical theories to be adapted or replaced and of models of historical development (e.g. models of social change). Alongside the empirical studies of regions, these activities provide fascinating methodological research frontiers for the decade ahead.

References

Blaut, J.M. (1962) 'Object and Relationship', *The Professional Geographer* 14 (6):1-7.

Braudel, F. (1979) *Civilization and Capitalism, 15th-18th Century: The Perspective of the World*, vol. 3, London.

Brunn, S.D. and Williams, J.F. (1983) *Cities of the World, World Regional Development*, New York: Harper & Row.

Buttimer, A. (1971) *Society and Milieu in the French Geographical Tradition*, Chicago: Rand McNally & Co.

— and Claval, P. (1987) 'IGU Discussion on Geography Today, Yesterday and Tomorrow', *The Professional Geographer* 39, May(2):221-3.

Claval, P. (1984) 'France', pp.15-41 in R.J. Johnston and P. Claval (eds) *Geography Since the Second World War: An International Survey*, London.

— (1987) 'The Region as a Geographical, Economic and Cultural Concept', *International Social Science Journal* 112:159-72.

Club of Rome (1972) *The Limits to Growth*, New York: Universe Books.

Dicken, P. (1986) *Global Shift: Industrial Change in a Turbulent World*, London: Harper & Row.

Dumolard, P. (1980) 'Le Concept de Région, Ambiguïtés, Paradoxes ou Contradictions', in Analyse Régionale, *Travaux de L'Institut de Géographie de Reims*, 41-42:21-32.

Friedmann, J. (1986) 'The World City Hypothesis', *Development and Change* 17:69-83.

Grigg, D. (1965) 'The Logic of Regional Systems', *Annals of the Association of American Geographers* 55:465-91.

Grotewald, A. (1971) 'The Growth of Industrial Core Areas and Patterns of World Trade', *Annals of the Association of American Geographers* 61:361-70.

Hartshorne, R. (1946) *The Nature of Geography*, 2nd ed. Lancaster, Pa.

— (1959) *Perspective on the Nature of Geography* Chicago: Association of American Geographers.

Humbert, M. (1986) La Socio-Dynamic Industrialisante, une Approche de L'Industrialisation Fondée sur le Concept de Système Industriel Mondial, *Revue Tiers Monde*, vol. XXVII, 107:537-54.

Johnston, R.J. (1982) 'The Local State and the Judiciary: Institutions in American Suburbia', pp.255-88 in R. Flowerdew (ed.) *Institutions and Geographical Patterns*, London: Croom Helm.

— (1985) 'The World is our Oyster', pp.112-28 in R. King (ed.) *Geographical Futures*, Sheffield: The Geographical Association.

Kostrowicki, J. (1975) 'A Key-Concept: Spatial Organisation', *International Social Science Journal* 27:328-45.

Labasse, J. (1955) *Les Capiteaux et la Région, Etude Géographique: Essai sur le Commerce et la Circulation des Capiteaux dans la Région Lyonnaise*, Paris: A. Colin.

— (1972) *L'Espace Financier, Analyse Géographique*, Paris: A. Colin.

Lichtenberger, E. (1984) 'The German-Speaking Countries', pp.156-84 in R.J. Johnston and P. Claval (eds) *Geography Since the Second World War: An International Survey*, London: Croom Helm.

Martin, J.P. and Nonn, H. (1980) 'La Notion d'Intégration Régionale', in *Analyse Régionale, Travaux de L'Institut de Géographie de Reims*, 41-42:33-48.

Pickvance, C.G. (1986) 'Comparative Urban Analysis and Assumptions about Causality', *International Journal of Urban and Regional Research*, vol. 10, 152-84.

Stewig, R. (ed.) (1979) *Probleme der Länderkunde*, Darmstadt.

Thomale, E. (1972) 'Sozialgeographie, eine Disziplingeschichtliche Untersuchung zur Entwicklung der Anthropogeographie' *Marburger Geographische Schriften*, vol. 5, Marburg/Lahn.

Thrift, N.J. (1986) 'The Geography of International Economic Disorder', pp.12-67 in R.J. Johnston and P.J. Taylor (eds) *A World in Crisis?: Geographical Perspectives*, Oxford: Basil Blackwell.

Whittlesey, D. (1954) 'The Regional Concept and the Regional Method', pp.19-68 in E.P. James and C.F. Jones (eds) *American Geography, Inventory and Prospect*, New York: Syracuse University Press.

Vidal de la Blache, Paul (1896) 'Principes de la géographie génerale', *Annales de Géographie* 5:129-42.

Vries, J. de (1984) *European Urbanization 1500-1800*, London: Methuen.

Wallerstein, I. (1974) *The Modern World-System: Capitalist Agriculture and the Origins of the World-Economy in the Sixteenth Century*, New York: Academic Press.

World Commission on Environment and Development (1987) *Our Common Future*, Oxford: Oxford University Press.

3 The changing view of the world in major geographical textbooks

Hans van Ginkel and Leo Paul

Geography has changed considerably since the Second World War. One of the most remarkable developments has been the decline of descriptive regional geography and the rise of new forms of applied geography, quite often quantitative in character. The changes in the content and orientation of geography are reflected in the changing popularity of major geographical textbooks which are included in the curricula of university geography departments. In this chapter we shall focus on the changing vision of the world as presented in the major geographical textbooks which have been used in the geography departments of Dutch universities since the early 1950s.

This review is unusual among studies in the field of the 'History of Geographical Thought' because we shall focus on change in views and ideas rather than on persons, schools of thought, or methodological questions. We are interested in the main line of thought running through much of the geographical work that has been published in the last decades, and its relation to the geographer's view of the world as it is reflected in diagrams and maps. For that reason it is important to take a closer look at the text and illustrations in major textbooks, because these can reveal changes both in geographer's world-views and in geography's paradigms.

The main thrust of this chapter is that, in the course of time, the geographer's view of the world changed considerably. It adjusted to change in geography's major applications and to its related changes in scope and methodology.

The world in the limelight

Before the Second World War, geography in The Netherlands was dominated by French descriptive regional geography. One of the most important textbooks at that time was Vidal de la

Blache's *Principes de Géographie Humaine*, which was post-humously published in 1922. Although in retrospect we can criticize this kind of geography, we must admire the way in which the 'founding father' of French descriptive geography and many of his students presented the results of research in maps. The book includes several maps of the world in colour, showing, for instance, the spread of population and the spread of plant and animal resources. The careful inductive way of working is well illustrated in Figure 3.1.

The way in which Vidal describes the relationship between culture and land use shows his cultural geographical approach. It is clear that the main object of Vidal's book was the study of the world, as the table of contents shows: First Part: The Spread of People on the Globe; Second Part: Forms of Civilization; Third Part: Circulation. The articulation of his view of the world was Vidal's main objective; the paradigm of *'genre de vie'* was his scaffolding.

Geography in The Netherlands in the early years after the war did not differ much from that in the pre-war period. The post-war economic reconstruction was not yet complete, and the 'you never had it so good' era was yet to come. The reconstruction of academic geography meant the reconfirmation of the pre-eminence of the French School. But the short period 1946-8, during which Jan Broek introduced cultural geography to Utrecht, was soon almost forgotten. French descriptive regional geographies, quite often almost literary pieces, dominated until the mid-1950s the list of required reading in the study of geography. Books such as Gourou's *Pays Tropicaux* (1947), Blanchard's *L'Homme et la Montagne* (1950), Gourou's *L'Asie* (1953), Derruau's *L'Europe* (1958), Cressey's *Asia's Lands and Peoples* (1945), Stamp's *Africa* (1953), and Gottman's *L'Amérique* (1949), demonstrate the image of geography at that time. Geography was the discipline that dealt with the earth and its major component parts. What else could be the subject of *geog*raphy, if not the *earth* itself?

But the French School soon went out of fashion. In the late 1950s a period of unprecedented economic prosperity was about to begin. At first, geographers focused their attention on the economic structure and growth of regions and, as a consequence, on interregional functions and relationships.

One of the first books to emphasize the economic structure of the world was Boesch's *Die Wirtschaftslandschaften der Erde*, published in 1947. The title of this book is characteristic of the transition of that period. The word 'landscape' was still in use, albeit with a focus on economic landscapes. But the world was still in the limelight. Figure 3.2, taken from

Figure 3.1 Autonomous developments of civilization, materials and building methods

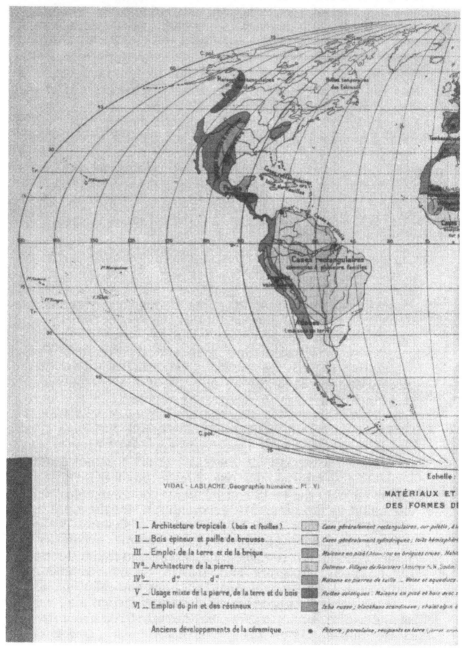

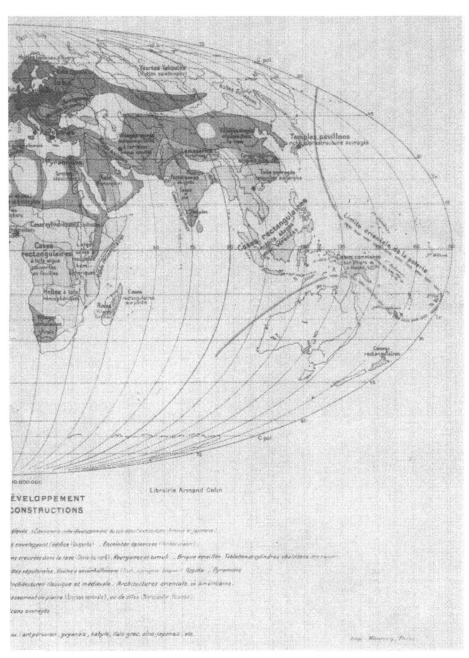

Source: Vidal de la Blache's *Principes de Géographie Humaine*, 1922, map 5

Figure 3.2 World economy and world trade

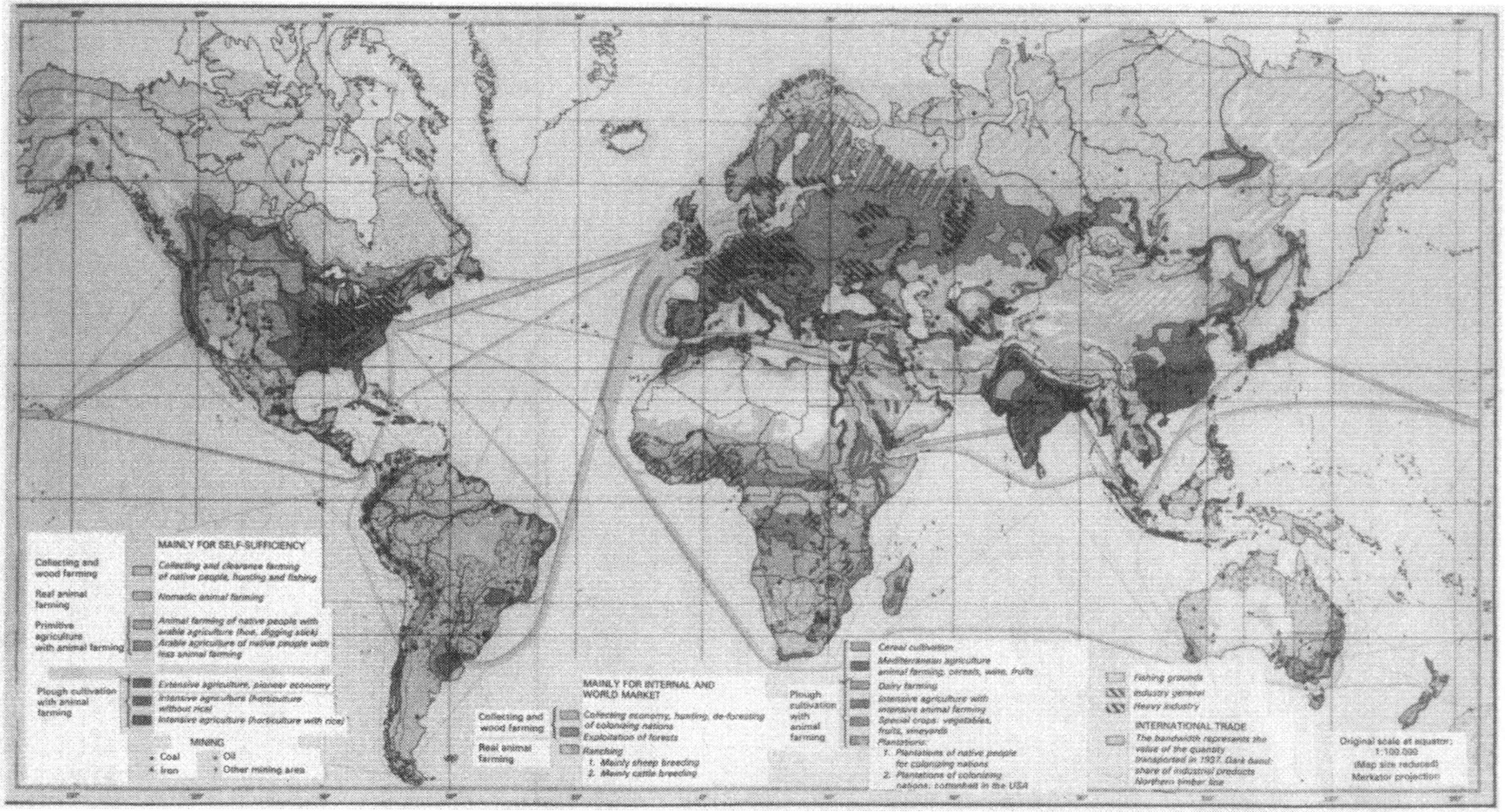

Source: Boesch, *Die Wirtschaftslandschaften der Erde*, 1947, appendage

Figure 3.3 Subsistence herding

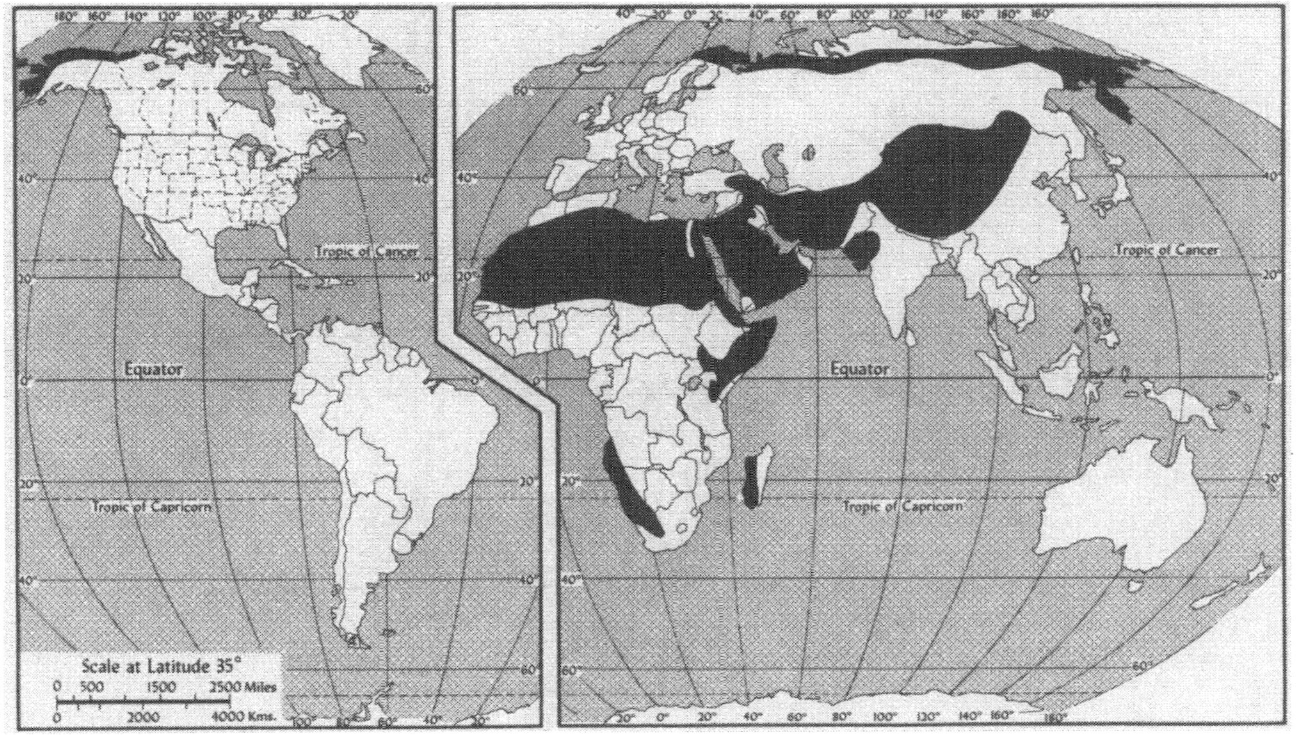

Source: Alexander, *Economic Geography*, 1963, p.35

Figure 3.4 Patterns of world occupance types

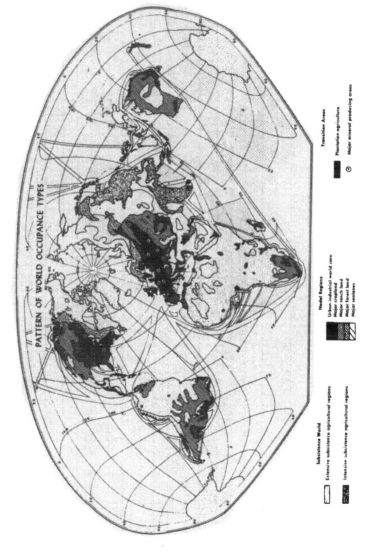

PATTERN OF WORLD OCCUPANCE TYPES

Subsistence World
- Extensive subsistence agricultural regions
- Intensive subsistence agricultural regions

Nodal Regions
- Urban-industrial world core
- Major cropland
- Major ranch land
- Major forest land
- Major sealanes

Transition Areas
- Plantation agriculture
- ⊙ Major mineral producing areas

Source: Alexander, Economic Geography, 1963, p.638

Boesch's book, shows that not only the regionalization of economic landscapes was important but also the trade relations between the regions.

Toward a systematic geography

The lists of required reading compiled by the Department of Geography at the University of Utrecht reveal that, since the late 1950s, textbooks on economic geography, such as those of Klimm *et al.* (1956) and later Alexander (1963), gradually became a common element in the curricula. In general these textbooks consisted of (a) an introduction to economic geography; (b) a treatise on natural resources and economic activities, in which almost every type of resource and activity was analysed and described separately; (c) a description of large-scale economic regions covering the whole world.

The pattern of the descriptions was different from that of the French School. The leading thread was economy, the language was English. The availability of data permitted much more quantification. In these textbooks physical environment and 'culture' are described only in relation to economic development:

> Cultural attributes differ from place to place as surely as do physical ones, and economic activity is strongly related to both.
>
> (Alexander 1963:11)

Most of the world maps in Alexander show the different types of economic activity and the spread of resources, each item projected separately on a global scale; for instance, the spread of subsistence herding (Figure 3.3). Alexander comments on his choice of cartographic representation:

> Throughout this book our attention has been drawn to numerous maps showing boundary lines around regions. Frequently the area inside the line has been solid black while outside the line has remained untouched. ... Such a device is helpful in presenting an initial view of spatial variation.
>
> (Alexander 1963:632)

Alexander does not enumerate all of the different economic activities just for fun. His ultimate objective is to combine all of the knowledge that he described before and to reduce it to two items which he presents at the end of his book in two maps of the world: one showing 'kinds and levels of economic development' and the other one showing the 'pattern of world occupance types' (Figure 3.4).

Alexander wants to show that this map has two messages: (a) the subsistence world is confined largely to the tropics and to very high latitudes; (b) the exchange world is basically a single far-flung area functioning around one great node, or core, that straddles the North Atlantic ocean (p. 637).

Although Alexander focused on economic activities, his ultimate goal was to describe spatial variations on a global scale. In the last chapter we read:

> Throughout this book our goal has been to understand spatial variation in economic activity around the world. As a means to that end, we have used the regional (or areal) method of analysis as a framework around which to organize knowledge of spatial differentiation. ... With all this as a background, we can now survey the totality of economic activity and make some concluding observations about the delimitation of regions, about general world regions of economic development and areal (regional) planning.
>
> (Alexander 1963:632)

It is important to emphasize that Alexander made use of his cartographic material as a means to reach his final geographical, not economic, objective. His kind of geography maintained its descriptive character: 'The geographical method of analysis enables us to understand more fully the world in which we live' (Alexander 1963:14).

We have already said that Alexander did not choose his working method just for the fun of it. But, as a result of this change in methodology, ordinary people lost their enjoyment of geography. For outsiders the discipline became less interesting. The mass of lay-geographers, who traditionally supported the discipline very strongly, became less interested as a consequence of the decreasing cultural and historical content of geographical studies. People fascinated by faraway regions, strange customs, and unknown natural phenomena gradually lost their interest in geography. Gradually, for geographers, the earth and its (major) constituent parts disappeared from the major textbooks.

Spatial analysis: the world disappears beyond the horizon

By the 1960s, geographers had focused their research more and more on quantitative analysis of socio-spatial processes and structures. Changes in science in general, such as the availability of new research methods and technical possibilities for data processing, and the strong tendency towards ever more precision, played a major role in this development. These scientific developments must be seen against the background of important changes in society. Because of the completion of the post-war reconstruction and the following period of unprecedented economic growth and — at least in The Netherlands — population growth, geographers focused their attention on issues of urban and regional planning. The growing awareness of the consequences of decolonization led to an increasing concentration on the problems and development of the former colonizing country itself. Besides, a growing trend can be observed within the welfare state towards social engineering.

The mainstream of spatial analysis reached The Netherlands only in the early 1970s. First of all, Haggett's *Geography: a Modern Synthesis* (1972) became popular, perhaps because this book tried to combine new methods of geographical research with the old traditions, as we can read in the introductory pages which were written by Meinig, the advisor on the book:

> Professor Haggett accepts the full tradition of geographer's work as worthy, insists that the essential questions geographers ask remain unaltered, and gives attention to all the grand themes of geographical inquiry; yet he casts these matters into a new mold, views them in fresh perspective, and brings to bear the full range of new working concepts and techniques.
>
> (Haggett 1972:x)

Haggett's own ideas are illustrated in the next pages:

> Geography is uniquely relevant to the current concern both with environment and ecology and with regional contrasts and imbalance in welfare ... Geographers are concerned with the structure and interaction of two major systems: the ecological system that links man and his environment, and the spatial system that links one region with another in a complex interchange of flows. ...

I have leaned towards systematic theory and hypothesis rather than elaborating many regional case studies. Although the emphasis throughout is on concepts and methods, it would be inconceivable to write an introduction to geography without numerous regional case studies. These range widely to stress the global variations in the environment and its exploitation as well as to illustrate differences in the temporal and spatial scales of operation of the forces discussed.

(Haggett 1972:xiv)

With Haggett's book, Dutch geographers could become used to the 'new' geography, often presented in a familiar global perspective. In his book Haggett used many cartographic illustrations, often creatively; see, for instance, Figure 3.5.

For this world map, Haggett used Paterson's index of potential productivity. The zones refer to potential plant growth estimated from climatic elements. The use of Paterson's index is pointless because its scale and values remain unclear. Notice that this mathematical approach of potential plant production has strong relations with Vidal's concept of *'genres de vie'*.

Later on in the 1970s, the world and its macro-regions gradually were given a more peripheral position in the study programme in universities and in the teacher training colleges: spatial theories and paradigms came into the limelight. This was clearly reflected in the required introductory reading lists at that time: textbooks such as Morrill's *Spatial Organization of Society* (1970), Cox's *Man, Location and Behaviour* (1972), and Lloyd and Dicken's *Location in Space* (1972). In these books, maps of the world are nearly absent. The world appeared only 'coincidentally' when spatial analysis took place on a global scale.

The world as a framework

By the second half of the 1970s, the situation was already changing again. Within geography the disillusion with the practical results of the dominant type of research led to a growing interest in the theoretical base of research designs as well as in theory construction as such. Efforts to improve the theoretical content of geography did not restrict themselves to 'theory and philosophy of geography', as had long been the case, nor to the incorporation of main ideas out of the philosophy of science in the discipline. Instead, they gradually

Figure 3.5 Productivity regions of the world

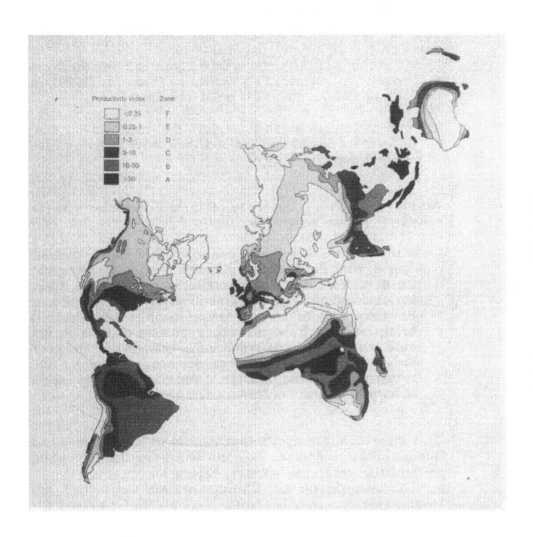

Source: Haggett, *Geography: a Modern Synthesis* 1972, p.34

focused on the linking of social science theory with so-called spatial theory. The growing interest in society for development issues and urban problems aimed the searchlight at: (a) the role of the state; (b) the way in which multinationals operate; (c) international - mainly economic and political - relations; (d) developments on a supra-national level.

An example of this trend is David Smith's *Where the Grass is Greener: Geographical Perspectives on Inequality* (1979). In the introduction he makes his point very clear:

> Ever since man began to explore the world around him, differences among peoples and their ways of living have aroused deep curiosity.
>
> (Smith 1979:i)

According to Smith, the study of these differences was the subject of human geography, but:

> Many of these differences are now a thing of the past, of course. With European colonization, empire-building and the development of an integrated world economy, there has emerged almost an international cultural ubiquity. ... Perception of differences among peoples and places is being replaced by a growing awareness that all do not experience to the same extent some common conception of the 'good life' that modern science, technology and industry have made possible. *Curiosity about differences is being replaced by concern about inequality.*
>
> (Smith 1979:15)

This introduction suggests that the world has regained the interest of the geographer. In Smith's book we can find several maps of the world; for instance, Figure 3.6.

This map dispels our illusion that the world itself has regained the attention. The world is merely the framework for the item that Smith puts at the centre, in this case inequality. For a geographer there is of course nothing wrong with examining inequality. But we must establish that, for Smith, the paradigm is chosen to fit the problem that he wants to solve, and this is reflected in his cartographic presentation of the world.

The late 1970s and early 1980s witnessed also a true revival of political geography (Taylor 1985; Johnston 1982; Claval 1984) as well as a revival of regional geography. Together these developments have led to a keen interest in world-systems approaches (Wallerstein 1979; Modelski 1978).

Figure 3.6 Where life is threatened: the world's hunger belt and major incidences of domestic violence

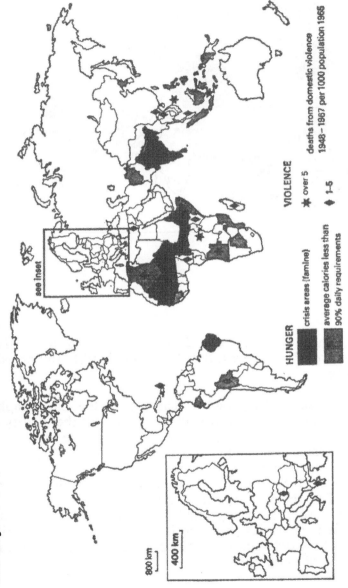

HUNGER

crisis areas (famine)

average calories less than 90% daily requirements

VIOLENCE

✳ over 5

◆ 1–5

deaths from domestic violence 1948–1967 per 1000 population 1965

see inset

800 km

400 km

Source: Smith, *Where the Grass is Greener: Geographical Perspectives on Inequality*, 1979, p.79

Figure 3.7 Employment in manufacturing by US-based multinational corporations in 1981

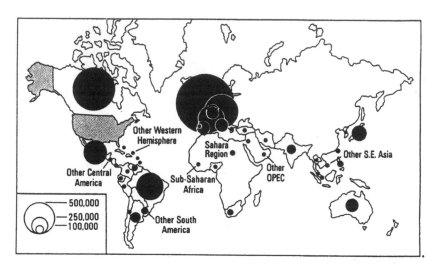

Source: Thrift (1983), reproduced in *A World in Crisis?: Geographical Perspectives*, R.J. Johnston and P.J. Taylor (eds) 1986, p.46

The disappointing fact remains that the cartographic representation of the world is often neglected; the quality of cartographic craftsmanship has become mediocre because of the developments described above. The subject of Figure 3.7, presented by Thrift in *A World in Crisis?: Geographical Perspectives*, 1986 shows the thematic interests of the early eighties. The theme is the most important aspect, the world serves as a sort of background. Almost all of Europe has disappeared from view because it is covered by dots. Which dot refers to which country is a puzzle that can be solved only by the best of geographers.

Conclusion

In this chapter we have tried to show that in the course of time the geographer's view of the world has changed considerably, as reflected in the graphic illustrations presented in the major geographical textbooks that are used in geography departments at Dutch universities. The maps of the world that Vidal de la Blache offered can be seen as the ultimate result of his careful inductive way of working. They offer an abundance of information and played an important role in his work. They were more than mere illustrations illuminating his book; Vidal wanted to show us his view of the world, using his paradigm as a scaffolding for that goal.

In the first decade after the Second World War, when changes in economy and society also affected the applications of geography, the world remained in the centre of geographers' work. The map we have shown from the book by Boesch illustrates that, for Boesch, as for Vidal de la Blache, the view of the world was the most important aspect of geography.

The economic geography of Alexander, for instance, must be characterized as a deviation from regional geography towards systematic geography (although at first the descriptive character remained). But in general the view of the world was not lost in Alexander's book. Through several thematic maps of the world, showing the spread of different economic activities, he ultimately combined these maps in two informative 'regional' maps of the world.

But with the scientific revolution in geography, which began in the 1960s, both the methodology and the view of the world in geography changed. During the heyday of spatial analysis, the world was no longer the major 'stage' for which the geographer wrote an interesting play. The world had been reduced to a 'scene', where the play was happening almost accidentally.

As a reaction to spatial analysis, a more 'problem-oriented' geography emerged, with more attention given to questions of spatial inequality, also on a global scale; a rehabilitation of the world and its macro-regions then seemed near. But if the cartographic representations of the world reflect that, we must conclude that this form of 'applied' geography also uses the world only as a spatial framework to illustrate a theme or a problem. In the maps that we have reviewed the world serves only as a large-scale background for thematic phenomena.

A 'new' regional geography should place the study of the world and its macro-regions back in the centre; this would have consequences for the cartographic presentation of the results of this renewed 'art' of geography. It is not enough to conclude that, in the making of cartographic material, major aspects of geographical craftsmanship have been lost in the course of time, as we have shown here. It is not only a question of cartographic techniques. It is especially important to realize that the *use* of maps of the world has changed. The first maps that we reviewed contained an abundance of information, which could be made available with the help of an extended legend. As geography developed and came to be taken seriously among the sciences, the maps of the world disappeared or changed considerably: too much reduction of information occurred, and legends were often missing or incomplete.

Now regional geography is making a comeback; it may not regain its position as 'the height' of scientific geography, but will certainly be accepted as a legitimate and important part of geography. In this respect, geography as a science has been able to reverse a trend toward decline. Until now this has not been noticeable in the use of cartographical representations of the world and macro-regions in new geographical textbooks. Regional geography is back on the 'stage' again, but in maps the world still serves only as backdrop.

References

Alexander, J.W. (1963) *Economic Geography*, New Jersey: Prentice-Hall.

Blanchard, R. (1950) *L'Homme et la Montagne*, Paris: A. Colin.

Boesch, H.H. (1947) *Die Wirtschaftslandschaften der Erde*, Zurich: Gutenberg.

Broek, J.O.M. and Webb, J.W. (1968) *A Geography of Mankind*, New York: McGraw-Hill.

Claval, P. (1984) *Géographie Humaine et Economique Contemporaine*, Paris: Presses Universitaires de France.

Cox, K.R. (1972) *Man, Location and Behaviour: an Introduction to Human Geography*, New York: Wiley.

Cressey, G.B. (1945) *Asia's Lands and Peoples: A Geography of One-Third of the Earth and Two-Thirds its People*, New York: McGraw-Hill.

Derruau, M. (1958) *L'Europe*, Paris: Hachette.

Gottmann, J. (1949) *L'Amérique*, Paris: Hachette.

Gourou, P. (1947) *Pays Tropicaux: Principe d'une Géographie Humaine et Economique*, Paris: Presses Universitaires de France.

— (1953) *L'Asie*, Paris: Hachette.

Haggett, P. (1972) *Geography: a Modern Synthesis*, New York: Harper & Row.

James, Preston E. (1949) *A Geography of Man*, Boston: Ginn.

Johnston, R.J. (1982) *Geography and the State. An Essay in Political Geography*, London: Macmillan.

— and P.J. Taylor (eds) (1986) *A World in Crisis?: Geographical Perspec-*

tives, Oxford: Basil Blackwell.

Klimm, L.E., Starkey, O.P., and Hall, N.F. (1956) *Introductory Economic Geography*, 3rd ed., New York: Harcourt, Brace.

Lloyd, P.E. and Dicken, P. (1972) *Location in Space: A Theoretical Approach to Economic Geography*, New York: Harper & Row.

Modelski, G. (1978) 'The Long Cycle of Global Politics and the Nation-State', *Comparative Studies in Society and History* 20:214-35.

Morrill, R.L. (1970) *The Spatial Organization of Society*, Belmont, NY: Wadsworth.

Smith, D.M. (1979) *Where the Grass is Greener: Geographical Perspectives on Inequality*, London: Croom Helm.

Stamp, L.D (1953) *Africa: a Study in Tropical Development*, New York: John Wiley.

Taylor, P.J. (1985) *Political Geography: World Economy, Nation-State and Locality*, London and New York: Longman.

Vidal de la Blache, P. (1922) *Principes de Géographie Humaine*, Paris: Armand Colin.

Wallerstein, I. (1979) *The Capitalist World-Economy*, Cambridge: Cambridge University Press.

4 Regions of the world system: between the general and the specific

Cees P. Terlouw

The world as a whole has become a prime area of study for the recently revived field of regional geography. Regions, irrespective of their scale, are no longer studied in isolation. The world is usually seen as the framework within which to study regions: it is 'our oyster' (Johnston 1985; see further: Buursink 1987; Pred 1984; De Pater and Van Ginkel 1987; Taylor 1985).

But how can the contextuality of the world over the regions be conceptualized? There are major differences of opinion about the way in which this global level is to be conceptualized. The ideas of Rostow (1961; 1978) lie at one extreme of the range of opinions. In his modernization theory, the world-economy is seen as a neutral meeting ground between states. He suggests that the outside world has very little influence over the internally generated development process in a state. The ideas of Wallerstein lie at the other extreme of the range of opinions about the way in which international dynamics are to be conceptualized. According to him, the linkages between states on a world level have become so great that the world as a whole is the only meaningful unit from which to analyse economic, social, and political dynamics. This world-system has a dynamic of its own and has an almost decisive influence over the dynamics of its constituent regions (Wallerstein 1974; 1979; 1980b; Wallerstein and Hopkins 1977).

In this chapter I shall discuss the way in which the world-system can be conceptualized. First the character of the world-system is explored. The following questions are raised. What are the boundaries of the world-system? Does it cover the entire globe, as Wallerstein claims, or are there several systems at a sub-global scale? Is the world-system basically an economic system, or do politics also play a decisive role? Finally, the spatial structure of such a world-system is analysed. This is initially done by following Wallerstein's theoretical description of the core, semiperiphery, and periphery. Then two impressionistic regionalizations of the world-system which were distilled from Wallerstein's work, are discussed. This will lead us to the important topic of mobility of states between different positions in the world-system. The conclusion will be that the dynamic world-system has not only a constraining but also an enabling influence on the development of a state.

One world-system or several?

Because Wallerstein's conception of the world as a single economic world-system is the most developed and the most extreme, I shall deal first with his approach. Then I will make a comparison with other authors who challenge the economic unity of the world-system. According to Braudel, there exist not one but several economic systems on a world scale. The economic nature of the world-system is also in dispute; Modelski especially distinguishes a separate political world-system with a dynamic of its own. The basic unit of analysis in the world-system theories is also disputed. Whereas Wallerstein, Modelski, and, to a lesser extent, Braudel use the state as their unit of analysis, others emphasize the role that world cities play in the present world.

The world as a single economic world-system

Wallerstein wants to delimit a meaningful context for the study of social change. This is difficult, because the boundaries of economic, social, cultural, and political processes do not overlap. A political system usually contains several different cultures. Economic exchanges often cross the boundaries of a political system. The simplest way to delimit a context for social analysis is to choose one of these dimensions of human behaviour.

   The choice of the political dimension is widespread in the social sciences. Generally the state is the implicit context for social analysis. But the influence of other states and the world-economy is so important for the developments in a state that a different context for social analysis must be chosen. Wallerstein chooses the economic process to delimit a social system, because this process has the biggest influence on the other aspects of human behaviour (Wallerstein and Hopkins 1977:114). But why choose just *one* aspect of human behaviour? As will be discussed below, it is more effective to use also political processes to bound the world-system.

So Wallerstein sees the world as a single world-system, based on economic processes. This was created in the sixteenth century in Europe and expanded until it encompassed the entire world at the turn of this century. According to Wallerstein, it was capitalist from the start. He defines capitalism as production for maximal profit in a market. This results in the endless accumulation of capital. Capitalism can flourish because of the political fragmentation of the world-economy. Although the degree of rivalry varies over time, no state is ever able to subjugate the entire capitalist world-economy. All states attempt to dominate the economy but, because they all try, no one succeeds. Therefore the market can operate quite freely in the capitalist world-economy (Wallerstein and Hopkins 1977:118-19; Wallerstein 1979:6, 15-16, 19, 66, 68, 120, 134, 147, 159, 272-3, 285).

The world as a collection of world-economies

But has there really been only one world-economy since the beginning of the sixteenth century? One can question both the spatial and the temporal boundaries of Wallerstein's world-system.

It is useful to make a distinction between a world-economy and the economy of the world (Braudel 1979). Although one can always speak of an economy of the world, the degree of cohesion varies. Today the economy of the world is very strong because the extensive network of trade relations has created virtually one world market, at least for some goods. The world-economy, on the other hand, 'only concerns a fragment of the world, an economically autonomous section of the planet able to provide for most of its own needs, a section to which its internal links give a certain organic unity' (Braudel 1979:22). Perhaps it is better not to speak of *one* world-system but to conceptualize the world as consisting of several different world-economies. These world-economies probably differ in their degree of cohesion and together they form a more loosely integrated world-system at the global level. This integration probably has a different character than integration at the level of the world economy. Today the integration in a world-economy may be primarily based on the integration of the production processes, whereas integration at the global level is probably based upon financial exchanges. Research into the different world-economies is needed to test these assumptions.

The temporal boundaries of Wallerstein's world-system are also debatable. Capitalist world-economies existed before the sixteenth century. In the eleventh century there were already signs of a world-economy in Europe. But this world-economy is not the same as the present one. Braudel depicts the history of the world-economy as a succession of world-economies. He connects this view with secular cycles, during which long periods of growth of a world-economy alternate with long periods of stagnation of that world-economy. During a period of stagnation a new world-economy arises out of the remnants of the old one. 'Crises: they mark the beginning of a process of destructuration: one coherent world system which has developed at a leisurely pace is going into or completing its decline, while another system is being born amid much hesitation and delay' (Braudel 1979:85).

What Wallerstein perceives as *one* capitalist world-economy is perceived by Braudel as a sequence of several different world-economies, with a different spatial structure and specific organizational characteristics. The location of the core changes between different world-economies (Table 4.1). The core regions differ, however, in their relative power over the world-economy. England was the most dominant and the most politically oriented. Genoa was the least conspicuous; its control over the world-economy was almost entirely based upon its dominant position in the credit system. It is also important to observe that Braudel sees the core of a world-economy as always centred in a world city. Only London and New York based their dominance upon their position in a productively superior territory. The other core cities established their position by controlling trade and capital relations (Braudel 1979:34-5, 153, 157, 210, 272, 295).

The world as political world-system

Not only the boundaries of the world-system are debatable; the criterion that Wallerstein uses to bound the world-system can also be questioned. As we have seen, Wallerstein chooses the simplest way to delimit a meaningful context of analysis, but this simplest way may not be the most accurate. The economic ties in the world are not the only important relations on a world scale (Modelski 1978). As Wallerstein himself stresses again and again, the creation of an interstate system was decisive for the creation of the capitalist world-economy (Terlouw 1985:16-27). He also acknowledges the existence of

Table 4.1 Wallerstein's hegemonic states, Modelski's global powers, and Braudel's core regions

Year	Wallerstein's hegemony	Modelski's cycle	Braudel's core	Year
1370			Venice .	1370
1490		‾‾‾‾		1490
1500		Italian wars	‾‾‾‾	1500
1510		‾‾‾‾		1510
1520			Antwerp	1520
1530		Portugal		1530
1540				1540
1550				1550
1560			(overlap)	1560
1570				1570
1580			Genoa	1580
1590		Spanish wars	‾‾‾‾	1590
1600				1600
1610		‾‾‾‾	(overlap)	1610
1620	‾‾‾‾			1620
1630	Holland	Holland	‾‾‾‾	1630
1640				1640
1650	‾‾‾‾			1650
1660			Amsterdam	1660
1670				1670
1680				1680
1690				1690
1700		Wars of Louis XIV		1700
1710				1710
1720				1720
1730		GB I		1730
1740				1740
1750				1750
1760				1760
1770				1770
1780			‾‾‾‾	1780
1790				1790
1800		Napoleonic wars	London	1800
1810	‾‾‾‾			1810
1820			(England)	1820
1830	GB	GB II		1830

Table 4.1 continued

Year	Wallerstein's hegemony	Modelski's cycle	Braudel's core	Year
1840				1840
1850	———			1850
1860				1860
1870				1870
1880				1880
1890				1890
1900				1900
1910		———		1910
1920		World War		1920
1930		I and II	———	1930
1940	———	———		1940
1950	US	US	New York	1950
1960				1960
1970	———		(US)	1970
1980				1980

a dynamic of the interstate relations which is at least partly autonomous from the dynamic of the economic world-system. But Wallerstein does not want to see this as constituting a separate world-system (Terlouw 1985:90-109). It is, however, better to conceptualize not only one or several economic world-systems but also at least one political world-system. It is obvious that a strong relationship exists between these two kinds of world-systems, but this relationship has to be empirically analysed and not taken for granted in the definition of a world-system as Wallerstein does.

The world consisting of actors other than the state

Other approaches to the study of integration at a global level question the basic unit of analysis. Whereas Wallerstein, Modelski, and, to a lesser extent, Braudel use the state as their unit of analysis, others emphasize the role that the world cities play in integration at a global level. Recently there has been a growth of interest in the global network of world cities (Cohen 1981; Friedmann 1986; Korff 1987). This world city network is largely seen as a network of control. World cities play an important role in the control over enterprises (transnational corporations) and the control over capital flows. World cities are important financial centres. They form units of analysis in the world-system at a level below the state.

The structure of the world-system

There is little disagreement between Braudel and Wallerstein about the way in which the capitalist world-economy functions. It operates through a geographical division of labour between rich and poor regions. The relation between the core and the periphery is one of exploitation. Goods that are manufactured in the periphery are exchanged for goods that are manufactured in the core, which are of a lesser value. The core states are able to enforce this unequal exchange because their state machineries are stronger.

The periphery and the core have also different economic and social structures. The economy of the core is much more diversified than the economy of the periphery. It uses the most advanced technology and has the most mechanized production process. The social relations in the core are also harmonious compared with the social relations in the periphery (Wallerstein 1979:38, 61, 97, 185, 274; Wallerstein 1974:355; Wallerstein 1980b:112-14, 284; Wallerstein and Hopkins 1977:129).

The semiperiphery occupies an intermediate position between the core and the periphery. In the semiperiphery there is a mix of peripheral and core activities. It exports peripheral goods to the core, and exports core goods to the periphery. The semiperiphery is being exploited by the core, but the semiperiphery exploits the periphery (Wallerstein and Hopkins 1977:128; Wallerstein 1979:39, 71-2, 97, 274).

With regard to the strength of the state, the semiperiphery also finds itself in an intermediate position. This has important consequences for the extent and the visibility of state intervention. The role of the state in the semiperiphery is very important. State intervention in the semiperiphery is very strong and very visible. The semiperiphery maximizes the need for and the possibility of intervening in the economy. The national economy is not strong enough to compete effectively in the world market, but the state is not too weak to let this pass. Semiperipheral states are therefore the most active states. As a consequence they are the principal sources of tensions in world politics (Wallerstein 1978:222; Wallerstein 1979:72, 274;

Wallerstein 1980b:113-14).

According to Wallerstein the semiperiphery is not a residue of states which is simply left after the identification of core and peripheral states. The semiperiphery has an important stabilizing influence on the world-economy. The semiperiphery depolarizes the relation between core and periphery. Because semiperipheral states profit from the world-economy, they have a vested interest in the endurance of this world-system. Not all exploited states will unite to overthrow the system, because the strongest among them — the semiperiphery — also profit from the exploitation of the periphery. The semiperiphery is also appeased by the possibility of gaining membership of the core (Wallerstein 1974:349-50; Wallerstein 1979:21-3, 34, 69, 233; Wallerstein and Hopkins 1977:129).

The dynamics of the world-system

The structure of a world-system is not static; it is subject to constant changes. First of all the degree of rivalry between the core states fluctuates over time. Sometimes a single core state is so strong that it dominates all other core states. States can also improve their position in the world-system by making use of the opportunities that the economic and political cycles of the world-system offer to states who want to improve their position in it.

Hegemony: the political cycle

Apart from core, semiperipheral, and peripheral states, there sometimes exists a still stronger state, the hegemonic state. Wallerstein characterizes hegemony as a sort of super core state. He refers to hegemony when a single core state is superior to all other core states in the capitalist world-economy. No other state nor coalition of states is able to threaten the superiority of the hegemonic state. Hegemony in a way leads to a fourth position in the world-economy but, contrary to the other positions, it does not always exist. It is a temporary condition of the capitalist world-economy: 'Hegemony therefore is not a state of being, but rather one end of a scale which describes the rivalry relations of great powers to each other' (Wallerstein 1984:39).

A state has achieved hegemony when it has reached dominance in all important economic domains. A hegemonic state has the most advanced production process, as a result of which its goods are competitive even on the home markets of its rivals. The hegemonic state is also the most efficient *trader*. It even has a share in the trade between other core states. The hegemonic state also has a superiority in the *financial* sector. It has the highest rate of return on capital, it lends money to others, and is an exporter of capital. As a result of its economic dominance, especially in trade, the hegemonic state also has a dominant position in world politics (Wallerstein and Hopkins 1977:121, 130; Wallerstein 1980b:38-9).

Modelski (1978; 1981; 1983) distinguishes a somewhat similar political cycle, during which relatively short periods of world power are alternated with periods of strife. His views differ from those of Wallerstein, however. Although Modelski (1981) links world power status with economic power, he defines world power status exclusively in terms of political strength in the global political system. He therefore characterizes the economically weak state of Portugal in the beginning of the sixteenth century as a world power (see Table 4.1).

Another difference between Modelski and Wallerstein is that the former's conception of the political cycle is more differentiated. Modelski distinguishes four stages of the political cycle: global war; world power; delegitimation; deconcentration. After the devastations of a global war, which may last for over a generation, there emerges a victorious world power, which is able to impose some sort of order on the global political system via a peace settlement.

> For the space of another generation that new power maintains basic order and is the mainspring of world institutions, often taking transnational forms. But the time comes when the energy that built this order begins to run down. ... Rivalries among the major powers grow fiercer and assume the characteristics of oligopolistic competition. Gradually, as order dissolves, the system moves toward its original point of departure.
>
> (Modelski 1978:271)

Table 4.1 also describes Modelski's periodization of the political cycle.

Braudel (1979) does not use the concept of hegemony, but his concept of the core is much more restricted than Wallerstein's. Braudel's definition of the core more or less matches Wallerstein's definition of hegemony. Table 4.1 therefore also shows the occupants of the core position according to Braudel.

Figure 4.1 The regions in Wallerstein's world-economy (*c.* 1900)

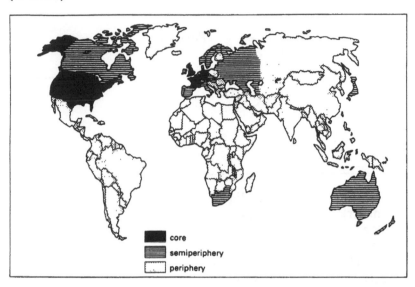

Source: According to Wallerstein as constructed by Terlouw

Figure 4.2 The regions in Wallerstein's world-economy (present)

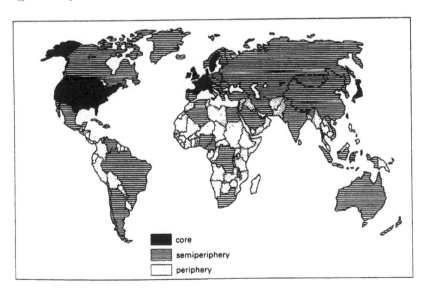

Source: According to Wallerstein as constructed by Terlouw

Table 4.2 Mobility in Wallerstein's world-system between 1900 (rows) and 1980 (columns)

	Core	Semi-periphery	Periphery	
C	FRG	GDR		
O*	France			
R*	UK			
E	US			
S	Belgium	Australia		
E	Denmark	Canada		
M	Ireland	Hungary		
I	Japan	* Italy		
P	Netherlands	New Zealand		
E	Austria	Norway		
R	Sweden	* Portugal		
I	Switzerland	USSR		
P		* Spain		
H		* Czechoslovakia		
E		South Africa		
R				
Y				
P		Albania	Afghanistan	Madagascar
E		Algeria	Angola	Malawi
R		Argentina	Bangladesh	Malaysia
I		Brazil	Benin	Mali
P		Bulgaria	* Bolivia	Morocco
H		Chile	Burkina Faso	Mauritania
E		China	Burma	Mozambique
R		Cuba	Burundi	Namibia
Y		Egypt	Central	* Nicaragua
		Finland	African Rep.	Niger
		Greece	* Columbia	North Yemen
		India	* Costa Rica	Pakistan
		Indonesia	* Dominican	* Panama
		Iran	Republic	Paraguay
		Israel	* Ecuador	* Peru
		Yugoslavia	* El Salvador	Philippines
			Ethiopia	Rwanda

Table 4.2 continued

Core	Semi periphery	Periphery	
	Lebanon	Ghana	Senegal
	Mexico	* Guatemala	Sierra Leone
	Mongolia	Guinea	Somalia
	Nigeria	* Haiti	Sri Lanka
	North	* Honduras	Sudan
	Korea	Iraq	Syria
	Poland	Ivory Coast	Tanzania
	Romania	* Jamaica	Thailand
	Saudi	Jordan	Togo
	Arabia	Kampuchea	Chad
	Taiwan	Cameroon	Tunisia
	Turkey	Kenya	Uganda
	Venezuela	Congo	* Uruguay
	Viet Nam	Laos	Zambia
	Zaire	Lesotho	Zimbabwe
	South Korea	Liberia	South Yemen
		Libya	

States with an asterisk before their name have belonged to the same region of the world-system since the seventeenth century.

The distribution of states over the regions of the world-system

One can also construct a more complete regionalization of the world-system. Figures 4.1 and 4.2 depict the regionalizations that were made by Wallerstein for *c.*1900 and *c.*1980 (Terlouw 1985:42-6). Table 4.2 presents this information in a different way. The asterisk before the name of a state indicates that it has belonged to the same region of the world-system since the seventeenth century. The trend that emerges from the maps and the table is one of upward mobility. Deterioration of the position in the world-system is extremely rare. In this century, only the position of the GDR in the world-system has deteriorated. This contrasts, however, with what Wallerstein says about the positions in the world-system. 'It does not follow that these zones are permanently immobile. Actors swap roles. What does *not* happen is that everyone becomes richer.

Only some do, and even then always at someone else's expense'
(Wallerstein 1980a:639).

Also remarkable is the diversity of the semiperiphery. It
consists of states which are reaching for core membership, of
recently promoted peripheral states, of states which structurally occupy this position, and of states which have dropped out
of the core. One can only distil this differentiation of the
semiperiphery from the work of Wallerstein. In his theoretical
essays, Wallerstein fails to differentiate between the semiperipheral states. One is drawn to the conclusion that he
considers only the mobile part of the semiperiphery, and
especially that part which aspires to core membership
(Wallerstein 1979:95-118). The semiperiphery is thus, contrary
to Wallerstein's assertion, a heterogeneous temporary station
between periphery and core rather than a homogeneous region
of the world-system.

Mobility between the regions of the world-system

How can this mobility between positions in the world-system
be explained? According to Wallerstein, internal processes play
an important role too. The strengthening of the state apparatus
is the key to an improved position in the world-system.
Intense state intervention in the economy is necessary to
stimulate national production and to protect the national
economy against incursions by the world-system. The national
bourgeoisie in (semi)peripheral regions is not strong enough to
compete effectively with the core producers. The bourgeoisie
is therefore dependent on the state for its economic survival
(Wallerstein 1979:72, 77, 85, 102-8).

But external constraints are of at least equal importance.
The penetrations of the world-system form the biggest
impediments for a state which wants to improve its position in
the world-system. There are several different ways in which
a state can limit the exploitation from abroad. One way is to
restrict contacts with foreign states. This policy of self-
reliance or autarchy is most forcefully pursued by the socialist
states. But the world-system is not only a constraining
influence on the development of a state. According to
Wallerstein, cycles in the world-system sometimes enable a
state, which has strengthened its state machinery, to improve
its position in the world-system. Primarily semiperipheral
states can periodically profit from economic stagnations in
the world-economy and from periods of increased rivalry
between core states.

Wallerstein's capitalist route towards a better position in the world-system

According to Wallerstein, the periodic economic stagnations have different effects in the three different regions of the world-system. Wallerstein distinguishes two periods of economic stagnation in the twentieth century: one between 1918 and 1939, and a second from 1967 onwards (Terlouw 1985:82-3). The semiperipheral states, because of their mix of core and peripheral activities, are almost ideally equipped to profit from an economic stagnation.

The conditions in the periphery deteriorate during an economic stagnation. This is the result of the habit of the core to attract to itself as much economic activity as possible during an economic stagnation. The production of some goods, which were formerly produced by the periphery and exported to the core, is shifted to the core. These industries are not as profitable as the usual core activities but, because profits are under pressure everywhere, the core producers use every profit source that they can get. The result is that the core is more self-sufficient *vis-à-vis* the periphery during a period of economic stagnation than during a period of economic growth. The periphery suffers the most from an economic stagnation (Wallerstein and Hopkins 1977:134; Wallerstein 1980b:129ff, 166).

The semiperiphery, however, can profit from an economic stagnation in the world-economy. The transfer of production to areas with low wages is another scheme of the world bourgeoisie to bolster its profit levels. During an economic stagnation, the semiperiphery is also able to obtain more concessions in their negotiations with the core. Overproduction makes the semiperiphery a more important outlet for the goods produced in the core. The semiperiphery can pick and choose to a certain degree between the different offers of the bourgeoisie in the core. The relative position of the semi-periphery as a whole *vis-à-vis* both the core and the periphery therefore improves during a period of economic stagnation. However, the core regains its share when economic growth returns. Only a very limited number of semiperipheral states, which have strengthened their state apparatus, are able to convert this temporary advantage into a permanent improvement of their position in the world-system (Wallerstein and Hopkins 1977:19, 76-9, 89, 97-100, 107-8, 247; Wallerstein and Hopkins 1977:129, 132, 134; Wallerstein 1980b:156, 179).

Wallerstein's Machiavellian route towards a better position in the world-system

Another way for a state to improve its position in the world-system is to make use of the periodic outbursts of political rivalry between core states.

During a period of hegemony, the world-economy operates most freely. Because of the economic superiority of a hegemonic state, it profits from the unrestrained functioning of the world market. A hegemonic state rarely intervenes in the functioning of the world market. The hegemonic state cracks down only on other states who try to compensate for the poor economic strength of their bourgeoisie by state intervention. A period of hegemony is therefore a period of free trade. During a period of hegemony, semiperipheral states have consequently little opportunity to improve their position in the world-system (Wallerstein and Hopkins 1977:131; Wallerstein 1980b:38, 61, 65, 269).

But hegemony is not stable. After several decades the other core states succeed in undermining the dominant position of the hegemonic state. A long period of rivalry between core states is the result. This rivalry enhances the possibilities of the other states to improve their position in the world-system. Because the core states are absorbed by the struggle amongst themselves, they are sometimes unable to stop a semiperipheral state from gaining membership of the core (Wallerstein 1980b:211, 241; Wallerstein 1979:99, 116).

It is difficult to determine the correctness of Wallerstein's ideas about the connection between the economic and the political cycles in the world-system, on the one hand, and the mobility of states between positions in the world-system, on the other. Only empirical research can answer this question. One has to consider when periods of economic stagnation and political rivalry occur in the world-system. It must be determined when given states improve their position in the world-system, and if this coincides with a period of economic stagnation and political rivalry. Furthermore, the mechanisms that Wallerstein uses to explain this mobility between positions in the world-system must be analysed. For instance, one has to look at the changing location of production in the world during a period of economic stagnation.

Conclusion

The most effective way to conceptualize the contextuality of the world over the regions is to see it as a relatively loosely integrated set of world-economies and a relatively separate political world-system. The precise ways in which these global systems are connected to each other must be subjected to empirical research. These systems are dynamic in nature because the position of states in the world-system is subject to constant change.

For regional geography, this means that, although the world is our oyster, the world is a constantly changing context. The external links of a region determine the structure of a region to a certain extent, but they also enable a region to improve itself. The section about mobility between positions in the world-system has shown that the world-system not only has a constraining influence over its parts but also sometimes enables a state to improve its position. This discussion has also led to the questioning of the correctness of Wallerstein's conceptualization of the structure of the world-system. It was shown that, contrary to Wallerstein's assertion, the semi-periphery was a rather heterogeneous group and that mobility in the world-system was almost exclusively upward.

Thus Wallerstein's world-system theory, however useful and stimulating, must be further elaborated. Braudel is right when he criticizes Wallerstein 'for letting the lines of his model get in the way of observing realities other than the economic order'. Wallerstein's approach is 'a little too systematic, perhaps, but has proved itself to be extremely stimulating. ... And it is this success that deserves most emphasis' (Braudel 1979 [3]:70).

References

Braudel, F. (1979) *The Perspective of the World: Civilization and Capitalism 15th-18th Century* (3 vols.). Harper & Row, New York.

Buursink, J. (1987) 'Regionale Geografie: Nieuw of Opnieuw?', *Geografisch Tijdschrift* 21:198-212.

Cohen, R.B. (1981) 'The New International Division of Labor: Multi-National Corporations and Urban Hierarchy', pp.287-315 in M. Dear and A.J. Scott (eds) *Urbanization and Urban Planning in Capitalist Society*, London: Methuen.

De Pater, B.C. and Van Ginkel, J.A. (1987) 'De Rehabilitatie van de Regionale Geografie', *Geografisch Tijdschrift* 21:253-57.

Friedmann, J. (1986) 'The World City Hypothesis', *Development and Change* 17:69-83.

Johnston, R.J. (1985) 'The World is our Oyster', pp.112-128 in R. King (ed.) *Geographical Futures*, The Geographical Association, Sheffield.

Korff, R. (1987) 'The World City Hypothesis: A Critique', *Development and Change* 17:483-95.

Modelski, G. (1978) 'The Long Cycle of Global Politics and the Nation-State', *Comparative Studies in Society and History* 20:214-35.

— (1981) 'Long Cycles, Kondratieffs, and Alternating Innovations: Implications for U.S. Foreign Policy', pp.63-83 in C.W. Kegley Jr and P. McGowan (eds) *The Political Economy of Foreign Policy Behavior*, Beverly Hills: Sage.

— (1983) 'Long Cycles of World Leadership', pp.115-39 in W.R. Thompson (ed.) *Contending Approaches to World System Analysis*, Beverly Hills: Sage.

Pred, A. (1984) 'Structuration, Biography Formation and Knowledge: Observations on Port Growth During the Late Mercantile Period, *Environment and Planning, D: Society and Space* 2:251-75.

Rostow, W.W. (1961) *The Stages of Economic Growth: A Non-Communist Manifesto*, Cambridge: Cambridge University Press

— (1978) *The World Economy: History and Prospect*, Austin: Macmillan Press.

Taylor, P.J. (1985) *Political Geography: World-Economy, Nation-State and Locality*, Longman, London.

Terlouw, C.P. (1985) *Het wereldsysteem: een interpretatie van het werk van I.M. Wallerstein*, Rotterdam: CASP.

Wallerstein, I. (1974) *The Modern World-System: Capitalist Agriculture and the Origins of the European World-Economy in the Sixteenth Century*, New York: Academic Press.

— (1978) 'World-System Analysis: Theoretical and Interpretative Issues', pp.219-35 in B.H. Kaplan (ed.) *Social Change in the Capitalist World Economy*, Beverly Hills: Sage.

— (1979) *The Capitalist World-Economy*, Cambridge: Cambridge University Press.

— (1980a) 'One Man's Meat', *Review* 3:631-40.

— (1980b) *The Modern World-System II: Mercantilism and the Consolidation of the European World-Economy 1600-1750*, New York: Academic Press.

— (1984) *The Politics of the World-Economy: The States, the Movements and the Civilizations*, Cambridge: Cambridge University Press.

— and Hopkins, T.K. (1977) 'Patterns of Development of the Modern World-System', *Review* 1:111-45.

5 Re-thinking regions: some preliminary considerations on regions and social change

Ray Hudson

Introduction

In this chapter I want to consider, in a preliminary fashion, the links between regional uniqueness, regional change, and social change and the ways in which human geographers and other social scientists have analysed them. For many years, one of the main concerns of human geography — indeed arguably its dominant pre-occupation — was the study of regions. Some geographers even went so far as to claim that there was a regional method which defined the discipline as a distinctive area of intellectual inquiry. The main focus of such studies was upon describing the unique characteristics of regions, a perfectly respectable though intellectually limited objective. In so far as there was any attempt to explain these, it turned inwards and drew upon the internal characteristics of the regions, the existence of which the analysis presupposed. Moreover, and this is especially clear in the work of De la Blache and his followers, there was a suggestion that regions reflected the emergence of an equilibrium between people and nature in a particular location that was, in the final analysis, unique to each region. A region's uniqueness might alter, slowly, over time but only as a result of the changing relationships between people and nature within it. There was little attempt to relate the internal characteristics of regions to wider social processes, to relate regional patterns to social processes, or to relate regional change to social change.

The emphasis upon the uniqueness of regions helped to trigger a counter-revolution which redefined geography as spatial science. In this view, geography should be directed towards the search for general laws of spatial structure or pattern, deploying the positivistic approaches of mainstream physical science. The focus of attention switched from the uniqueness of place to the generality of space, often reduced to transport costs. In so far as a concern for regions remained, it was in terms of classificatory categories which were to be found at a fairly early stage on the route from description to explanation. It rapidly became clear, however, that simply producing generalizations about spatial pattern was a very limited exercise and that explaining regularities in spatial pattern necessitated their relation to social processes. In general, however, the relations between spatial pattern and social process were conceptualized as static and cross-sectional around some assumed equilibrium state; a very partial and restrictive viewpoint. The mechanisms whereby one equilibrium state was replaced by another remained unexplored territory, for example. In this sense the dynamics of spatial change could not be analysed. There was, though, one important exception to this general tendency towards ahistoricism, originating in the diffusion studies of Hägerstrand which were addressed precisely to the issue of explaining how spatial patterns changed over time. They represent one of the very few attempts, located within a definition of geography as spatial science, which explicitly sought to theorize about spatial change; they are of particular interest here in so far as they tried to specify the social processes underlying spatial change, the links between the two, and the spatial unevenness in the resultant diffusion patterns. It is for this reason that they are considered at greater length in the next section, which seeks to identify their strengths and weaknesses in understanding the links between social change, regional change, and regional uniqueness.

In more recent years, as part of a widespread and varied reaction against the excesses of spatial science, there has been a considerable re-evaluation of the relationships between society and space. One strand of this has been the development of a humanistic geography which, in its exploration of the meanings that places have for people, revives in a rather different way concerns that were central to regional geography. But, like the regional geography of earlier eras, it remains essentially descriptive in its approach. Another strand, which has emerged from an engagement between human geography and modern social theory, puts much greater emphasis upon explanation and upon explicating the inter-relationships between the uniqueness of place and more general social processes. The key proposition of such an approach, in the present context, is that regional uniqueness reflects the interplay of broad social forces in and with the specificities of regions. This necessitates the specification of the key dimensions of a given society — principally class relations but also gender relations, relations of ethnicity, and so on, and the articulations between these — and of how they combine in particular ways in particular places to produce unique regions. At the same time, however, this relationship between regional uniqueness and social processes is reciprocal, so that the former influences the development of the latter. This type of approach therefore allows the links between social change, regional change, and regional uniqueness to be explored from a different perspective. This is attempted, tentatively and schematically, in the third section of the chapter.

Geography, diffusion studies, and social change

Several years ago, Sarre (1972:50) claimed that 'the most highly developed area of geographic enquiry into the development of processes in space and time is that dealing with the diffusion of innovations'. At the time, this was quite a reasonable claim because human geographers had mostly neglected the dynamics of changing spatial patterns and it is for that reason that diffusion studies form the focus of this section. I am concerned not to review the literature on diffusion studies comprehensively but rather to examine it very selectively in terms of how such studies conceptualize (often implicitly) the processes of social change that are reflected in the changing spatial patterns of innovation diffusion.

However, this raises the issue of what is meant by 'social change'. At one level, social change can be thought of in broad macroscopic terms as the change from one dominant mode of social organization to another: from feudalism to capitalism, for example. Such changes, though revolutionary in their content and impacts, are often not evolutionary in the time period over which they occur. They are, however, all-embracing in their effects, permeating all spheres of life: economic, social, cultural, political. As an integral part of this process, they involve the creation of new spatial patterns; literally new geographies. By and large, this sort of social change has been generally neglected by diffusion studies that are cast in the spatial science mould (though for partial exceptions see Gould 1970; Leinbach 1972; Soja 1968). At another level, social change can be thought of in terms of changes within the parameters that define a society as, say, feudal or capitalist; most of the diffusion studies in geography are of this sort.

The main emphasis in diffusion studies in geography lies not with the social processes of change, however, but with the spatial pattern of innovation diffusion. Two main spatial regularities in innovation diffusion have been noted: contagious and hierarchical. These are essentially analytic distinctions, since in practice most spatial diffusion patterns contain elements of both. How, though, are the processes that give rise to these observed spatial regularities conceptualized? In the case of contagious diffusion patterns, which are associated above all with Hägerstrand's pioneering studies (for example, see Hägerstrand 1967), the process is conceived in terms of face-to-face contact between individuals; information and personal contacts are prioritized. It is easy to point out that in the modern world most information is received via the mass media but this is not the fundamental point of criticism. This lies in the over-emphasis upon the role of information, the limited conceptualization of information, and the stress upon asocial individuals in conceptualizing the processes involved; crucially, people are conceived as atomized individuals, with no sense of the structural relations in which they are embedded, or of the relations between them as agents and these structures. Consequently, it is impossible to obtain any firm purchase on the links between social process and spatial pattern and change.

At first sight, a more promising approach is offered by those diffusion studies which emphasize the roles of urban hierarchies and both public and private corporate organizations in channelling information flows and innovation diffusion down hierarchical structures. Certainly, this gives a more reasonable representation of the spatial pattern of diffusion in many areas (for example, see Berry 1972). Even so, the explanatory promise is often mainly illusory, since the existence of these hierarchies is largely taken as given; for example, there is no recognition of the sense in which urban hierarchies themselves were often produced as one element in quite profound social changes. Gould (1970), though he makes some penetrating points in this regard, does not develop them. Again, there is a tendency to ignore the historical trajectory along which multinational conglomerate companies travelled to reach their present complex and hierarchically ordered internal organizational structures (though links to notions such as product life cycles and changing spatial divisions of labour could be easily enough made). Similarly, there is scant reference to the evolution, organization, and functions of modern capitalist states that have brought them to the stage when they too have a complex, internally differentiated, and hierarchical organization. Furthermore, the purposes for which and the social processes by which information flows are created and transmitted through such organizations are largely ignored. The treatment of information remains at best partial, even although it is in some ways more satisfactory than in the Hägerstrand school of contagious innovation studies.

Others have developed much fuller critiques of diffusion studies (for example, Blaikie 1978; Gregory 1985), but the crucial point in terms of the argument to be developed here is that such studies are woefully weak in their conceptualization of the processes of social change. If anything, there is a tendency to reduce the processes of social change to the spatial pattern of innovation diffusion (for example, see Berry 1978). As such, they offer no real basis from which to investigate the relationships between social change and regional change. Indeed, if Sarre's (1972) claim is taken seriously, they point to the massive barriers that are created by the spatial science approach in general towards the investigation of the relationships between these changes. Diffusion studies, which are set in this mould, tend to be built around an assumption that the norm is for everyone everywhere to adopt innovations; non-adopting areas constitute some sort of anomaly. Put another way, regional specificity becomes a deviation from an assumed normal pattern of social change; or, as Relph (1976) might have put it, non-adopters are places that refuse to become placeless.

Social theory, social change, and regional uneven development

In the last decade or so there has been a quite radical reassessment of the conceptualization of regions as one element in a broader exploration of relationships between society and space. This has involved moving away from a concern with developing classificatory categories of places (such as cities or regions) to one that centres upon relationships between social structures and the spatial patterning of societies; in short, to a concern with what Soja (1983) has termed the 'spatiality of society'. This focus upon the 'spatiality of societies' has been one product of a convergence of interests between some social theorists and some human geographers which stresses the point that places are socially produced; it is in this context that Massey (1978) was insistent that regions must emerge from analysis (or maybe synthesis: see Massey and Allen 1985) rather than be presupposed by it. It also follows that if regions are to be understood as being socially produced, they must be understood not simply as areas of space but as distinctive time-space domains (Giddens 1985). In other words, their definition requires not simply the demarcation of lines on the ground but the periodization of social change. This allows regions to be defined, as it were, as discrete and distinctive socially produced time-space envelopes This implies a need for a methodology which allows the mechanisms to be uncovered that are involved in bringing about changes in societies which are constituted, as a condition of their existence, in spatially differentiated patterns. Regions are and must be 'chaotic categories'; that is, regions are objects which 'cut across many structures and causal groups in a "chaotic" fashion. It is not always possible or desirable to reduce the object so that it is less chaotic, because it may nevertheless be of interest as a whole' (Sayer 1984:227). This suggests the need for an approach which allows for the disentanglement of causal structural processes from contingent factors. It is in this context that Massey and Allen (1985) argue the case for synthesis as a method for unravelling the processes that are involved in the social production of regions as unique places which are linked into chains of unequal interdependencies.

What is involved here is essentially a recognition that: (a) regions (places) are socially produced and reproduced and, in the last analysis, are unique; (b) unique regions (places) are linked via social relationships into chains of unequal interdependencies; (c) broader processes of social change are in turn and in part shaped by the specificities of unique regions (places).

This points, therefore, to a reciprocal relationship between more general social processes and the social production/reproduction of places as unique. The reason for this is the interrelationships between the existing natural and socially produced attributes of these places with the broader processes of economic, political, and social change, many of which are increasingly operative not just at the national but at the global scale. Regional uniqueness and regional patterning, then, are not simply to be read off mechanistically as products of social process; quite the contrary, because as well as regions, one might say, helping to create themselves, these broader social processes are both reproduced through and changed by the specificities of regions. One very clear example of this reciprocal relationship can be seen in the interrelationships between changing forms of corporate organization, the changing geography of production, and the characteristics of different places in the construction of new spatial divisions of labour, initially between regions within the boundaries of one national state but increasingly now recast on the global scale.

For example, Nissan, a major Japanese producer of vehicles may have been persuaded to locate an assembly plant in north-east England in the mid-1980s because of the attractions that the region offered to it in terms of profitable production. The region is within the European Community's external tariff wall and so provides access to a major market; any new project within the region was eligible for major state grants to help to meet fixed capital investment costs; the region offered an abundant supply of pliable labour from which Nissan could carefully select its workforce. As a result, Nissan's location there set in motion changes which are beginning to rework the region's specificity. At the same time, however, Nissan's locational decision has triggered processes of change which will reverberate throughout the rest of western Europe's car-producing regions. One such change is the redefining of labour productivity norms as a result of the company's being able to recruit a young, physically fit workforce, committed to the company. Put another way, social change occurs both in and through places. It occurs unevenly over space, to be sure, but such unevenness is now understood very differently to the way in which it would be by diffusion modellers who would see it in terms of anomalous non-adopting regions failing to conform to some (maybe only implicit) normative model of social change.

How are the relationships between social structure, social change, and regional change most appropriately to be understood? They are certainly not to be regarded as deterministic, in the sense of being able to deduce regional patterning from the central structural relations around which societies revolve and through which they are defined; to be able to specify a society as feudal, capitalist, or state socialist does not allow too much to be said about how it will be patterned over space. Clearly the links between regional patterning and social structural relations are contingent and this is central to the way in which regional specificity is conceptualized above. The same causal mechanisms and the same general social processes can lead to different forms of regional stasis or change, depending upon the pre-existing characteristics of regions and it is for this reason that regions are and must be 'chaotic conceptions'.

How are the mechanisms to be revealed through which structural relations with causal powers combine together with contingent factors to produce particular patterns of regional differentiation and change? Indeed, how are they to be conceptualized? The answer to this partly depends on the level at which social change is being conceptualized. For example, at a very broad macro-scale it is clear that the relationships between social structure and spatial patterning will tend to differ between feudal and advanced capitalist societies. This is a difference that will reflect decisive changes in people's capacity to modify the natural environment, in production, transport, and communications technologies, and so on, as well as in the changes in social relationships in which these technological developments are embedded and of which they form a part (for example, see Toft Jensen *et al*. 1983). Even so, the spatial patterning of advanced capitalist societies will in all probability include relics of the spatial patterning of feudal societies. These traces of, for example, land ownership patterns and the built environment both survive and help to shape, though certainly not to determine, the transition from one set of dominant social relationships to another.

Similarly, within one particular mode of production or, more accurately, a society that is dominated but not exclusively occupied by one decisive set of social relationships, there will be definite, though still contingent and by no means unchanging, relationships between social change and regional differentiation. Perhaps the simplest way to develop this point is by reference to a specific example: regional differentiation within capitalist societies. At one level, there are several well-known statements that attempt to generalize about this; for example, in terms of core and periphery. But the point that I want to emphasize here is that core and periphery are not static but relational categories: cores are and can only be defined in relation to peripheries, and vice versa. Furthermore, such a dichotomy begs crucial questions, such as: how do cores and peripheries come to be socially produced? Why do some regions change, as they undoubtedly do, from core to peripheral status or from periphery to core?

At a very abstract level, in terms of a value theory analysis of the capitalist mode of production (CMP), it can be convincingly demonstrated that capitalist development is an inherently uneven process and part of this unevenness involves regional differentiation. Harvey (1982) has produced an unrivalled restatement of Marxian political economy which situates regional differentiation within a more general value theory analysis of the uneven development of capitalism. In this sense, he powerfully demonstrates the way in which regional differentiation is inscribed in the inner logic of the CMP. This is undeniable but it is also necessary to go beyond it, as he himself has emphasized. Analysing the CMP at this level of abstraction cannot reveal which regions will become cores, or centres of accumulation, and which will become peripheral at particular phases of history; once again, regionalization necessitates the defining of boundaries in both space and time. Certainly one can go on to periodize capitalist development in terms of broad epochs, such as the transition from competitive capitalism to monopoly capitalism; or more recently, in the vocabulary of regimes of accumulation, in terms of a transition from Fordism to a flexible regime of accumulation. Whatever the vocabulary and categories used, however, the point remains that each of these qualitatively different phases of capitalist development involves, as part of its characteristic and defining ensemble of social relations, a distinctive geography (albeit one that emerges from and is laid down upon, rather than one that wholly replaces, the geography of the previous phase(s), although the extent to which traces of previous patterns remain is itself variable). The point emerges nicely in a recent paper by Harvey. Referring to the new regime of flexible accumulation, he points out that it 'is marked by a startling flexibility with respect to labour processes, labour markets, products, and patterns of consumption. It has, at the same time, entrained rapid shifts in the patterning of uneven development, both between sectors and geographical regions (Harvey 1987). Even so, suggestive though such comments are, the mechanisms that link spatial patterning and the causally determining powers of the decisive social relationships still require tighter specification.

How, therefore, are the links between these social structural relationships and contingently produced regional patterns to be conceptualized and understood? To put the point slightly differently, there are important questions to do with how the structural relationships of and limits to capitalist societies — as opposed to the more abstract notion of the CMP — are socially produced and reproduced. For example, how the antagonistic class structural relationship between capital and labour, around which surplus value production revolves, is produced is clearly a crucial issue that needs to be explained rather than assumed away; likewise, the issue of how competition between capitals over realized surplus value is resolved needs to be explained rather than assumed away. It was precisely these sorts of issues that began to command greater attention as part of a more generalized critique of structuralist Marxism (for example, Althusser 1977; Althusser and Balibar 1970) that emerged in structurationist approaches in the 1970s (for example, see Giddens 1978). Moreover, the convergent emphasis in the work of social theorists, such as Giddens (1981; 1984) and Urry (1981a; 1981b), and geographers, such as Gregory (1978; 1981; 1982), Thrift

(1983), and Soja (1980; 1983), in seeking to demonstrate the importance of place — alongside dimensions such as class, gender, and race — in the constitution of society, provided valuable insights into the links between social structure, social change, and regional change. This involves interpolating what might be described as 'intermediate level' theoretical concepts between the high level abstractions of value theory and the empirical evidence of regional differentiation, via seeking to uncover the reciprocal relations between structures and agents in societies that are constituted both in and as specific time-space regions.

Some insights into these relationships can be gleaned from concepts that are central to Hägerstrand's time geography (for example, Hägerstrand 1970). In this it is recognized that the reproduction of social life depends upon knowledgeable human subjects tracing out routinized paths over space and through time, engaged in projects in conditions that are shaped by the interplay of capability, coupling, and authority constraints. The limitations to approaching these relationships of agency and structure from this perspective are considerable, however, and revolve around the point that the links to the social processes through which these time-space constraints are produced remain at best partially specified. Furthermore, focusing upon established routine patterns of behaviour tends to beg questions about how these routines were socially established as part of a taken-for-granted pattern of everyday life and, indeed, how deroutinization and the transition to new forms of routine behaviour become established. In this sense, the extent to which the issues of social change can be addressed from this time-geographical perspective are limited.

Another, more promising approach (although one that in turn draws upon Hägerstrand's ideas) to beginning to unpack the relationships between social structure and spatial pattern, by analysing the links between agents, structures, and the production of places, is to be found in Giddens's structuration theory (see Giddens 1981; 1984). Whilst this has its problems (for example, see Gregson 1986), it offers valuable insights via its insistence upon recognizing that agents both are shaped by and help to shape structural relations; what Giddens refers to as the 'duality of structure'. But it also involves acknowledging that what agents do partly is shaped by and partly helps to shape where they live, learn, and work. There is, however, considerable debate surrounding the concept of agency. For Hägerstrand, agents are and can only be individuals. Although Giddens too lays considerable stress upon individuals as agents, he also acknowledges the importance of the institutional settings in which they act and, in some senses, he regards institutions and other forms of social collectivities as possessing powers of agency. Giddens, however, decisively rejects the notion that 'classes' can act as 'collective agents'. It seems to me that from one viewpoint it is possible, but not particularly helpful, to claim this. Seen from other viewpoints, though, such a strong assumption cannot be sustained. While capital and labour clearly do not act in and for themselves as unified classes (bourgeoisie and working class), based on a shared understanding of their class interests and class structural position, this is no more than a recognition of the fact that capitalist societies are much more complicated than the highly abstract two-class representations of the CMP that Marx used in unravelling the fundamental inner structural core of the social relationships of capitalism. It is equally clear that there are shared (though shifting) collectivities of interest which allow us, at a lower level of abstraction, to talk of the strategies of capital and labour, typically mediated by the capitalist state, in relation to the social production and reproduction of regions as part of the overall process of capitalist development. (For the moment, I want to set aside the issue of divisions within the classes of capital and labour and between and within capitalist states: these are considered below.) In other words, the trajectory of social development is most appropriately conceptualized as being decisively, though by no means entirely, shaped both by and through the development of class struggle, conflict, and compromise. At the same time, it must be recognized that Marx's very abstract two-class model was intended not to portray the real historical processes of class formation and conflict but to provide the theoretical basis for investigating these concretely in specific historical-geographical contexts.

It is in this light that I want to consider some possibilities regarding agreement or disagreement between capital, labour, and the state as to the desirability or otherwise of regional change. First, however, it is important to recognize that there is a presupposition here that capitalist development has proceeded sufficiently for a distinctive regional patterning to have emerged, which is different to but overlaid upon that of the previous pre-capitalist phase(s). This initial regional patterning of capitalism itself was a product of conflict and struggle as the old feudal order was replaced by the new capitalist one, a process that itself involved momentous changes in the composition of both dominant and dominated classes. Clearly, as capitalist social relations began to be established, new cities, towns, and regions were constructed as an integral part of the process of capitalist development. Capitalism both presupposed and created this new geography as an integral part of establishing its hegemonic sway within the new social order. In this sense, capitalism developed its own distinctive spatiality as an integral part of the process by which it became established and expanded. Accumulation depended upon profitable production and this required *inter alia* the construction of new factories and mines, new transport systems, and new residential areas. These last were to provide housing for the migrants who flocked to these areas to provide the crucial commodity of labour-power, as the new class relations between capital and labour were sedimented into place. All this occurred in what had previously been feudalistic agricultural areas, now irrevocably transformed as the new core regions of capitalist production were created in them. The new social order, based on the hegemonic dominance of capitalist relations of production, and the new regional patterning went hand in hand, inseparably linked. These newly created regions, the birthplaces of industrial capitalism, were built around deep class divisions, which often erupted into bitter conflict. At the same time, however, the regions developed a sense of economic and social cohesion as interrelated capitalist conglomerates emerged, which were in a sense committed to producing in them and, in so doing, provided a source of waged employment for a working class that had little choice but to sell its labour power somewhere. With subsequent developments in the capitalist economy, new regional patterns were created as the locational requirements of profitable production altered. However, the general point that I want to establish is that, in the initial creation of these regional patterns, the power of capital, and, to a generally lesser and more variable degree, that of the state, was more or less unfettered by consideration of the interests of labour. Once regions were constructed, however, the interests of labour as well as those of capital and the state became represented in the processes that were intended to generate regional change or to preserve regional stasis.

Some of these possibilities regarding stasis or change and the class interests favouring them are set out schematically in very simplified form in Table 5.1. What I want to do is briefly to sketch out some of the relationships that could exist between regional change and social change; what I will not do is to produce detailed historical-geographical exemplars for each of the four cells of this matrix. We begin, then, from the point that there is already a socially produced regional pattern which the interests of capital, labour, and the state might want either to preserve or to alter. Again, it is important to stress that, whilst the capacity to pursue these interests is, as it were, structurally loaded in one direction, since capital is in a qualitatively different position to labour in terms of its powers and resources to pursue its interests, it does not follow that change or stasis is always as capital desires. This is partly because of the role of the state (though there are other crucial reasons that are touched upon below).

Table 5.1 Regional change and class interests: a simple typology

	Agreement; consensus	*Disagreement; conflict*
Change the region	Capital, labour, and the state agree	Capital, labour and the state disagree
Stabilize the region	Capital, labour, and the state agree	Capital, labour, and the state disagree

Although in the last analysis the state is the guarantor of capitalist social relations, it is for this very reason (among others) that in a democratic society the state cannot simply be seen passively to reflect the interests of capital all of the time, since the state is subject to political pressure from labour, not least via the electoral process. This is suggestive of the crucial mediating role that the state can come to take in formulating policies to change or to ossify regional patterns. Regions, which are certainly 'chaotically conceived' in Sayer's terms, none the less become the objects of state (regional) policies but the content of these policies depends in part upon the relative strengths of capital and labour in influencing their formation and implementation. This again introduces an element of contingency into the form and direction of regional change.

There is, nevertheless, a crucial qualitative difference between those situations where there is a consensus between capital, labour, and the state as to the desirability or otherwise of regional change (whether with reference to the entire regional patterning or more usually to a selected region) and those situations where there is disagreement and conflict. The politics of regional attachment and class allegiance can clearly become very complicated in such situations.

Consider first those cases where there is consensus as to the desirability of preserving a particular regional pattern, or of preserving the specificity of particular regions. There were several cases in the UK in the 1950s, for example, in which a powerful consensus emerged among the decisive centres of power in the state, capital, and labour to prevent change in regions such as north-east England and south Wales: the state was concerned to guarantee cheap energy from coal; capital could see the prospects of healthy profits from producing steel, ships, etc; labour could anticipate jobs and wages in these industries, in strong contrast to the 1930s. Little more than a decade later, in large part because of changes in international energy markets and the international division of labour in industry, these same areas were characterized by the emergence of a consensus between capital, labour, and the state as to the necessity for regional modernization and change. Sometimes this was linked into corporatist projects for modernization of national economies. (Despite this consensus, however, the resultant regional modernization programmes were only very partially implemented as intended. This raises important questions about the limits to state power and about the determinants of the direction of regional change.)

Those situations where there is conflict between capital, labour and the state about the direction of regional change are in general more complicated, however, as class interests and regional attachments begin to diverge. In these situations, the asymmetry of power relations becomes crucial. Disinvestment decisions by capital, as part of a global strategy to boost profits, may be resisted by labour, generally without success. This leads to demands for alternative jobs to which the state cannot or will not respond. If the state does respond with new policy initiatives, capital refuses to invest, or to invest on a sufficient scale to create alternative jobs, because other places offer more profitable locations. Such a situation has recurred regularly in the old industrial regions of the advanced capitalist world over the last decade; former core regions of capitalist production have become deindustrialized as coal-mines, steelworks, shipyards, and so on have shut and state reindustrialization policies have signally failed to produce their intended effects. Clearly in those situations the permutations of possible forms of conflict and cooperation are much greater. However, the point remains that, in situations of conflict, capital is generally in an infinitely more powerful position to pursue its interests than is labour, while growing internationalization of production has altered the balance of power between capital and many national states.

This is a further complication that has been alluded to but now must be more explicitly recognized. Moving to a lower level of abstraction, it is necessary to acknowledge that there are important differences in interests within each of capital, labour, and the state. Capital is not homogeneous in its interests, there are differences between, say, finance and manufacturing capital, between capital in chemicals and in steel, and between competing chemical companies such as Bayer and ICI. Indeed, inter-capitalist competition is central to the motor of capitalist development. In like manner, labour is not homogeneous in its interests, being divided by industry, occupation, etc. and often along dimensions such as gender and race. Furthermore, capitalist states are not monolithic juggernauts relentlessly pursuing generally agreed objectives but are themselves characterized by different objectives; for example, between central government departments. Moreover, there is a strong territorial element running through these divisions within capital, labour, and states. Most obviously, this is visible in the constitution of capitalist states as national states but there are also intra-state divisions between central and local government, whilst both capital and labour are chronically divided on a territorial basis. Given that the interests of different fractions of capital and labour are routinely, in part, defined territorially and that states are defined in terms of national territories and are organized on a sub-national territorial basis, then it is clear that questions of social change, which revolve around class conflict (without denying the relevance of dimensions such as gender and race) are inextricably bound up with those of regional change, and vice versa.

Conclusions

Not so long ago, the concern of 'traditional' geographers with describing the uniqueness of regions seemed to be something of an embarrassment to their more 'modern' colleagues who were seeking to produce general theory about spatial structure. Yet, in the last decade or so, knowledge of the specificities and uniqueness of regions has again re-emerged as an issue of central importance, and not only to geographers. Once again, questions of regional division have assumed a central position on the political agenda in countries such as the UK. Not so long ago, it would have been very difficult to envisage that the concerns of 'traditional' regional geographers and those of grand social theorists would converge and, at least to a degree, come together. Yet, in the last decade or so, this is precisely what has happened. As a result, questions about the relation-ships between regional uniqueness and social structure, regional change, and social change have begun to be addressed seriously. As increasing attention has been directed towards recognizing the spatiality of societies, regional uniqueness has come to be seen not only as an integral and constitutive element of broader social relationships but also as something that is most appropriately understood in terms of synthesis within the region of place-specific characteristics with these more generally established social processes.

Such a perspective places considerable demands upon sound, theoretically informed empirical research in elucidating the links between regions and societies. Regions certainly need to be understood in the context of theories about social structure and change but, conversely, the particular regional patterning of a given society is a contingent matter, a product of the interrelationships between general structural relations, which are endowed with causal powers, and the specificities of places. In this sense, the 'traditional' object of analysis of geographers now holds a central place in modern social theory. There is, however, a sting in the tail. The way in which regions are now conceptualized is far removed from the approach of traditional regional geography and it is not without irony that reasserting the central importance of regions exposes still further the limitations of both regional geography and geography as spatial science. Seen from a different perspective, however, the recognition that incor-porating a sense of spatiality is vital in understanding social structure and change indicates a re-evaluation of disciplinary boundaries which must extend, of necessity, beyond geography to the rest of the social sciences and which opens up exciting possibilities in terms of the study of regions.

References

Althusser, L. (1977) *For Marx*, London: New Left Books.
— and Balibar, E. (1970) *Reading Capital*, London: New Left Books.
Berry, B.J.L. (1972) 'Hierarchical Diffusion: The Basis of Developmental Filtering and Spread in a System of Growth Centres', pp.108-38 in Hansen, N. (ed.) *Growth Centres in Regional Economic Development*, New York: The Free Press.
— (1978) 'Geographical Theories of Social Change', pp.17-36 in Berry, B.J.L. (ed) *The Nature of Change in Geographical Ideas*, Illinois:

Northwestern University Press.

Blaikie, P. (1978) 'The Theory of Spatial Diffusion of Innovations: A Spacious Cul-De-Sac', *Progress in Human Geography* 2:268-95.

Giddens, A. (1978) *Current Problems in Sociological Theory*, London: Macmillan.

— (1981) *A Contemporary Critique of Historical Materialism*, London: Macmillan.

— (1984) *The Constitution of Society*, London: Polity Press.

— (1985) 'Space, Time and Regionalization', pp.265-95 in Gregory, D. and Urry, J. (eds) *Social Relations and Spatial Structures*, London: Macmillan.

Gould, P.R. (1970) 'Tanzania 1920-63: The Spatial Impress of the Modernization Process', *World Politics* 22:149-70.

Gregory, D. (1978) *Science, Ideology and Human Geography*, London: Hutchinson.

— (1981) 'Human Agency and Human Geography', *Transactions, Institute of British Geographers*, NS6:1-18.

— (1982) *Regional Transformation and Industrial Revolution*, London: Macmillan.

— (1985) 'Suspended Animation: The Statis of Diffusion Theory', in Gregory, D. and Urry, J. (eds.) *Social Relations and Spatial Structures*, London: Macmillan, pp.296-336.

Gregson, N. (1986) 'On Duality and Dualism: The Case of Structuration and Time Geography', *Progress in Human Geography*, 10, 2, pp.184-205.

Hägerstrand, T. (1967) *Innovation Diffusion as a Spatial Process*, Chicago: University of Chicago Press.

— (1970) 'What about People in Regional Science?', *Papers of the Regional Science Association* 24:7-21.

Harvey, D. (1982) *The Limits to Capital*, Oxford: Basil Blackwell.

— (1987) 'Flexible Accumulation through Urbanization: Reflections on "Post Modernism" in the American City', Paper presented to a Symposium at Yale School of Architecture, 6-7 February 1987.

Leinbach, T.R. (1972) 'The Spread of Modernization in Malaya, 1895-1965', *Tijdschrift voor Economische en Sociale Geografie* 63:262-77.

Massey, D. (1978) 'Regionalism: A Review', *Capital and Class* 6.

— and Allen, J. (1985) *Geography Matters!*, Cambridge: Cambridge University Press.

Relph, E. (1976) *Place and Placelessness*, London: Pion.

Sarre, P. (1972) 'Diffusion', pp.44-67 in *Channels of Synthesis: Perception and Diffusion*, Units 16-17, Block 5, New Trends in Geography, Milton Keynes: Open University Press.

Sayer, A. (1984) *Method in Social Science*, London: Hutchinson.

Soja, E. (1968) *The Geography of Modernization in Kenya: A Spatial Analysis of Social, Economical and Political Change*, Department of Geography, University of Syracuse Geographical Series No. 2, Syracuse, New York.

— (1980) 'The Socio-Spatial Dialectic', *Annals of the Association of American Geographers* 70:207-25.

— (1983) *The Spatiality of Social Life: Towards a Transformative Retheorization*, Mimeo, Los Angeles, UCLA.

Thrift, N. (1983) 'On the Determination of Social Action in Space and Time', *Environment and Planning D: Society and Space* 1(1):23-57.

Toft Jensen, H., Hansen, P.A., and Serin, G. (1983) 'Capitalist Technology and the Change of the Labour Process', pp.89-103 in A. Gillespie (ed.) *London Papers in Regional Science* 12, London: Pion.

Urry, J. (1981a) 'Localities, Regions and Social Class', *International Journal of Urban and Regional Research* 5:455-73.

— (1981b) *The Anatomy of Capitalist Societies*, London: Macmillan.

6 What about regional geography after structuration theory?

Joost Hauer

Since the early 1980s, social science has witnessed a remarkable convergence in theory development. The pioneering work of the human geographers Van Paassen and Hägerstrand, the historian Braudel, and the sociologist Giddens has made it clear that a fruitful social theory should be integrated within a time-space perspective. Social scientists should be concerned with the structure and evolution of society in time and space. These insights are gradually undermining the deeply rooted presupposition that principles of spatial ordering are at the basis of the spatial arrangement of mundane, regional, and local societies. It is becoming increasingly clear that nation-states, cities, municipalities, and so on are social categories which are based on principles of social ordering. Of course, space plays a role, but its effect may only be understood in connection with the human and the social. Social theory has been enriched with a new perspective in which the dialectical relation between societal structures and individual human agents is the central issue.

In this chapter I summarize these developments and illustrate their fundamental importance for human geography. Some old problems will be reassessed in the light of this new perspective. I shall discuss some of their consequences for empirical research and the state of regional geography.

Towards a reconstruction of social theory: the basis of structuration theory

Van Paassen's view: human geography is a social science

To show how scientists from diverse social science disciplines led the way to a convergence of social theory would go beyond the scope of this chapter. I prefer to draw attention to an eminent Dutch geographer who is relatively unknown internationally but who had a remarkable foresight of coming events in theory development.

Christiaan van Paassen, the Nestor of Dutch theoretical geography, reviewed developments in human geography in his farewell lecture on the occasion of his retirement in September 1982. By that time, the spatial analysis tradition had passed its peak of international acceptance. The critical arrows emanating from (neo)marxism and philosophy of science had reached their target and the behavioural, humanistic, and phenomenological orientations were reaching maturity. The basic point that Van Paassen made with reference to these criticisms is that, although the critics were essentially right, their arguments were only partial; much better arguments are — and always have been — available in the history of geography itself. No one who is acquainted with Van Paassen's work would have expected otherwise (for sources of his work in English, see Van Paassen 1976; 1981).

In his lecture, Van Paassen (1982:37-43) focused attention on three issues: the formal-spatial approach, systems analysis, and the neglect of regional geography. In the effort to formalize space, abstract concepts such as position, direction, and dispersal became of central importance. These concepts have no connection with substantive reality. As a distinction is made between formal and real spaces, social relations become spatial relations and the spatial organization of society becomes subject to investigation. But Van Paassen argues that spatial configurations are at best a means by which to study societal problems. The ordering principles in society are social and not spatial. Geographers should study how societal relations change and integrate or disintegrate in time and space. Society, as it manifests itself, materializes in space and time. The resulting variety is what the geographer should study.

The second point addressed by Van Paassen concerns the systems approach. This approach is based on the assumptions that a territorial society has a specific internal order and can be differentiated from other societies. The internal structure can be taken as the point of departure in an explanation of why it functions as it does. However, this starting-point implies the selection of characteristics on the basis of their coherence, while the existence of coherence itself still remains to be demonstrated. In the complex reality of society, this implies a selection in which only subsystems or partial wholes are studied or else only one specific level of analysis is chosen. In both cases, essential relations are omitted and the integration of social, economic, political, and cultural phenomena within a specific spatial context becomes obscured.

Here Van Paassen raises the issue of the neglect of regional geography. Although, he argues, geographers should study territorial integration and the ensuing variability, most prefer to study regularities instead of variations. The fundamental task of geographical endeavour, however, is to study variety and to explain the impact of communication of individual and social actors, their common action, their solidarity, confrontation, production, consumption, and creation in territorial coexistence.

It is quite clear that Van Paassen tries to confront newly formulated problems with the traditional basic attainments of the geographical discipline. In fact he argues that human geography is essentially a social science with respect to the type of explanations that we look for; although the orientation is spatial, the explanations should be social. In this sense the parallel with Braudel's work in the field of history is evident: time is the historian's orientation, but frequency, duration, and sequence are not the ordering elements in history (Van Paassen 1982:38). There is also a striking parallel with recent work in psychology (see Peeters and Monks 1986).

Giddens's view: social science is also a spatial science

It is interesting to see how Giddens adopts this train of thought and his theory of structuration. His conception of time and space is fully in line with that formulated by Van Paassen and Braudel. A remarkable convergence of theoretical developments is emerging, especially where it concerns the crucial role of time and space in the reproduction of everyday life. Social theory in Giddens' view is not only social as such but also inherently historical and geographical.

In his book, *The Constitution of Society* (1984), Giddens gave the most complete overview of his ideas to date. I shall not discuss his theoretical work in detail here but will only summarize his findings and then elaborate on some specific points. In addition to Giddens's book, other sources for this summary include Spaargaren *et al.* (1986), Munters *et al.* (1985), Thrift (1983), Moos and Dear (1986), and Gregory and Urry (1985).

One of the fundamental contributions of Giddens to the development of social theory is his ability to bridge the gap between theories about human agents and theories about institutions, or, as it is also put, between voluntarism and determinism, or subjectivity and objectivity. In order to do this, Giddens reconceptualizes the notions of agency and structure, enhancing their dual nature: the one presupposes the other. With regard to agency, human actions are either routine or explicitly intentional, but they are always contextual. If they are routine actions, the human agent almost automatically knows what to do in a certain situation. These actions are not thought about and are not performed as a result of a choice from a set of possible alternatives. The human agent acts according to available practical knowledge, which is everything that is known and that has to be known in order to be able to function in the immediate contexts of social life. This knowledge is available, though not necessarily conscious. Think, for example, of the way in which we speak our native tongue, using the grammatical rules of that language without being able to reproduce or write down those rules. This practical knowledge exists independent of the acting person. If human actions are explicitly intentional, the reasons for such behaviour can be given by the agent, who has thought about it and has made a deliberate choice on a discursive level of thinking in order to reach a certain goal. In this case the actions are also contextual, of course, because they are deliberately performed within a specific situation. The possibility of unconscious behaviour will be disregarded for the moment.

Given this combination of practical and discursive knowledge, human actors are 'knowledgeable and capable agents', having both the necessary knowledge and the knowledge of what they do and why (although aware that this knowledge is incomplete). This highlights the difference from deterministic and objectivistic interpretations of reality.

Regarding the reconceptualization of the concept of structure, Giddens makes a distinction between structure and system. He argues that all actions in society, all social practices, comprise a set of events which is reflected upon and steered by actors; they do not form a random set of events. The mere fact that these real practices actually occur means that they are situated in space and time or, as Giddens puts it, 'stretched across time and space'. What we do in a certain place and at a certain time is understandable only because social rules and patterns of behaviour connect these actions with what happened at the same place yesterday or at the same time somewhere else. Social interaction through time and space has a systematic character. Daily life consists of temporal and spatial routines, spatio-temporal activity patterns, in terms of which our activities are organized. The concept of system is now described as the real activities of agents as they are situated in time-space. The concept refers to the relations between individuals and groups. Social systems are reproduced relations that stretch across space and time so that the practices being realized in the present are connected with practices that have already been or are going to be realized.

In contrast to the concept of system, the concept of structure refers to something that exists outside time and space. Structure is described as the recursively organized rules and resources which are used by human agents in the context of their daily lives and which are reproduced at the moment that they draw upon them. Here the dual meaning of structure comes to light: structures are both the medium and the outcome of the situated practices that make up the system. As Moos and Dear put it:

> Structure is the medium whereby the social system affects individual action and the medium whereby individual action affects the social system. The outcome of these individual-system interactions always (in varying degrees) affects the structural rules governing the next interaction. Thus, the theoretical separation of structure and system enables Giddens to capture both agency and structure in the production and reproduction of social life without according primacy to either.
>
> (Moos and Dear 1986:234)

The societal (inter)action through which this happens is called structuration. Social systems exist through structuration, as the outcome of the acts of human agents. In other words, structuration refers to the structuring of social relations across space and time within the context of the dual relation between agency and structure.

Three components are distinguished in the order to which structure refers. All three are inherent to social interaction: signification, domination, and legitimation. Structures of signification become manifest through the communication of meaning. Actors draw upon rules of interpretation to be able to give significance to interaction. At the same time, applications of these rules means reproduction of the structures of significance. Structures of domination refer to the asymmetrical distribution of resources in interaction. Using power in interaction does, of course, in return lead to a reproduction of the existing asymmetry. Legitimation then refers to the application of norms in legitimizing behaviour.

From this discussion it becomes evident that the two concepts of agency and structure presuppose each other logically. When agents act, they draw upon structural properties that allow actualization; at the same time, the structural properties are being reproduced. It is essential to note here that, since human agents do not have complete knowledge about their actions, we have to reckon with unacknowledged conditions and unintended consequences of actions. The process of motivation, rationalization, and reflexive monitoring of action is understandable only if we realize that social action is bounded by these unacknowledged conditions and unintended outcomes. Moreover, we should realize that the latter two stand in a reflexive relation to one another: unintended outcomes, in so far as they have a systematic effect on social reproduction, become conditions for action as well.

This will suffice as a summary of the basic elements of structuration theory. Most writers who, to date, have summarized, interpreted, or criticized Giddens agree to a large extent. However, opinions diverge when discussions reach the point where agency and structure have to be connected more explicitly with time, space, institutions, region, and other concepts.

Space, time, and social theory

From the previous discussion we may conclude that, according to Giddens, it is precisely the duality of structure that connects the production of social interaction and the reproduction of social systems across space and time (Moos and Dear 1986:234). It is interesting to note that Thrift (1983) poses the question of whether we need one or several mediating concepts between agency and structure. In comparing his interpretation of the answers to this question that are given by Bhaskar, Giddens, Bourdieu, and Layder, Thrift implicitly draws the conclusion that a mediating concept is needed. At the same time, he ascertains that the answers that are given differ from author to author. Despite the fact that Thrift uses a diagram wherein mediating concepts are formulated between structure and practices, while the question was about the structure and agency, he concludes that in Giddens's formulation of structuration theory the mediating concepts between structure and practices are system and institutions. This, however, does not correspond to the above conclusion. I much prefer the way in which Moos and Dear (1986:241) answer the question. They present structure as the concept between system and agents which characterizes their inter-relationship.

To see why this is a more fruitful way of looking at things, I shall now discuss the importance of time and space and make clear how institutions may still be taken into consideration.

The treatment of time and space

In structuration theory, time and space are not just the dimensions of the context within which the human agent acts, they are essential parts of the design of actions. What is relevant is not only the packing problem (i.e. how to use time and space and how to decide about priorities in sequencing and space-use) but also the time and space perspective that is connected with every action, and, above all, the way in which action here and now relates to other actions (earlier, later, at the same place, or somewhere else). This is the reason why several central concepts in structuration theory are defined by using space-time categories.

> Structure refers to the structuring properties allowing the 'binding' of time-space in social systems, the properties which make it possible for discernably similar social practices to exist across varying spans of time and space and which lend them 'systematic' form. To say that structure is a 'virtual order' of transformative relations means that social systems, as reproduced social practices, do not have 'structures' but rather exhibit 'structural properties' and that structure exists, as time-space presence, only in its instantiations in such practices and as memory traces orienting the conduct of knowledgeable human agents. This does not prevent us from conceiving of structural properties as hierarchically organized in terms of the time-space extension of the practices they recursively organize. The most deeply embedded structural properties, implicated in the reproduction of societal totalities, I call structural principles. Those practices which have the greatest time-space extension within such totalities can be referred to as institutions.
>
> (Giddens 1984:17)

Institutions, be they institutionalized behaviour or organizations, therefore refer to standardized or more enduring actions in time-space. To understand how institutions develop and show continuity, it is essential to understand how action taking place at a specific time and place relates to institutionalized forms of action.

The treatment of time

Giddens sees time as occurring at three levels (Giddens 1984; Munters *et al.* 1985). *Dasein* is the temporality of the human life cycle; *durée* is the temporality of immediate experience, of the continuous activities and interactions of daily life; finally, the *longue durée* refers to institutions or, more generally, standardized forms of behaviour (across generations). These three aspects or levels of temporality are closely interwoven. Every moment of social interaction is part of the life cycle and at the same time connects with the continued existence of institutions. It is again the duality of structure that binds the reproduction of the *durée* and the *longue durée* of institutions and structural properties (Moos and Dear 1986:237). What is essential is the presence of differing time perspectives, ranging from the temporality of the moment to the longest time perspectives in which the deeply embedded structures slowly change (over ages). Giddens's use of the terms *Dasein* and the Braudelian *longue durée* is rather confusing when one compares their use with the original connotation.

The treatment of space: the locale

The parallel between the treatment of time and space becomes more evident when we discuss the two main concepts that were developed with respect to space: locale and presence-availability.

> Locales refer to the use of space to provide the settings for interaction, the settings of interaction in turn being essential to specifying its contextuality.
> (Giddens 1984:118)

Locales may be designated by their physical properties (physical elements and artefacts), but essentially the concept refers to the way in which the spatial context is part of the interaction. Locales may range from dwellings to factories, to small towns and cities, to nation-states, and to the world. Most locales are internally regionalized in a specific manner; these regions within locales are of crucial importance, as they constitute contexts of interaction (Giddens 1984:118). This regionalization of locales has to do with the zoning of space and time in relation to social practices. Certain parts of time and/or space are reserved for specific purposes (activities or individuals); compare, for example, the interactions inside the rooms of a dwelling. Localizing and regionalizing in time and space are important structuring elements of everyday life.

The scales at which locales and their regionalizations exist and change make it possible to initiate research into how practices, social systems, and institutions stretch across time and space without a priori ideas about systems and societies. As Giddens noted in an interview: 'What a society is has to be directly analysed' (Gregory 1984:127). The way in which time-space is connected with social systems differs from case to case. This is clarified by the concept of presence-availability. Small-scale communities are generally characterized by high presence-availability. For interaction to take place, only small distances in time and space have to be bridged. Interaction partners are usually directly available and most interaction has a face-to-face character. With the development of modern media, interaction over long distances grew more complex. Instead of going into this theme in greater depth, however, I turn to Thrift's treatment of the locale.

Thrift (1983:40) starts his discussion by stating that any region provides the opportunity for action and the constraints upon action. Note that with the term region he refers to a number of different but connected settings for interaction; he adopts the term locale to transmit this meaning. Then he argues that a particular pattern of production (and consumption) results in a particular pattern of locales. Furthermore he posits that certain of these locales will be dominant in the sense that time has to be allocated to them. These dominant locales are said to have five effects (Thrift 1983:40):

(a) They structure people's life paths in space and time;
(b) They influence life paths through induced constraints on interaction possibilities;
(c) They provide the main areas (in time terms) within which interaction with other people takes place, experiences are being gathered, and a common awareness is engendered;
(d) They provide the activity structure of day-to-day routines;
(e) They are the major sites of the process of socialization that takes place from birth to death, within which collective modes of behaviour are constantly being negotiated and rules are learned and created.

To some extent these five effects summarize aspects of the previous discussion. But from the way in which they are formulated, it seems that too much emphasis is laid upon the constraint side. According to Thrift, these 'dominant locales provide the most direct link between interaction structure and objective social structure, because they are the sites of class production and reproduction (Thrift 1983:40).

Thrift also argues that some locales and institutions may be related to a dominant locale in some way that situates them in an historical pattern of determination (Thrift 1983:41). Without going into more detail with respect to his argument, we may conclude that Thrift's concept of structure is fundamentally different from that of Giddens (see also Moos and Dear 1986:240). The fact that the locale provides various opportunities for and constraints upon action is quite definitely something different from being the site of the determinate working of an objective social structure. The duality of structure as defined by Giddens is not fully adopted in Thrift's view.

On the other hand, the addition of the term dominant to the concept of locale, in those cases where the locale's constraining influence is relatively strong, seems to be useful and meaningful. In reality there is a continuously changing balance between enabling and constraining in the sense that the margins of constraints may widely differ. More important, perhaps, is that the functioning of locales and the internal regionalization of locales have to be connected with the time perspective of interaction. The degree of dominance of a locale is most probably strongly connected with properties that are embedded in the (*durée* and) *longue durée* of reproduction.

With regard to earlier observations about mediating concepts and the place of structure, it should now be clear that the duality of structure as defined in structuration theory can only mean that there is no place for either determinism or voluntarism. At the heart of the theory is the agent-structure-system interaction and the primacy or importance of any of these cannot possibly be specified a priori (Moos and Dear 1986:241). Moos and Dear's discussion of this problem is that:

> This implies that individuals retain the characteristics which enable them to alter the social system, at the same time acknowledging that the socialization of individuals through living in a social system is of crucial importance, and cannot be discerned outside a socially defined context. Furthermore, structures, although ontologically real, are not accorded primacy since they reflect the structural properties that are embedded in the *longue durée* of social reproduction. Structures only exist in the way in which agents draw upon and reproduce them in interaction.
>
> (Moos and Dear 1986:241)

Consequences for empirical research: the interpretative method and the human agent as level of analysis

The adoption of structuration theory has clear consequences for empirical research in human geography. This section comprises some general remarks; the next section is more specific regarding the position of regional geography.

Results from empirical research which is explicitly undertaken with a structurationalist view in mind are scarce, one of the most remarkable exceptions being Dear and Moos's analysis of the 'ghettoization' of ex-psychiatric patients in an urban context (Dear and Moos 1986). Although we cannot draw on a broad experience with the application of structuration theory, some general guidelines may be found which suggest how social research should be oriented. Giddens himself formulates several statements with respect to this matter. To summarize and review these, I draw from two sources: his book (Giddens 1984:281-354, especially 281-8) and a lecture that he presented in Wageningen in 1984 (Munters *et al*. 1985:48-56). From the central theorem of the duality of structure, the resulting interwovenness of agent and system, and the importance of the time-space constitution of social life, three points emerge that need specific attention (Giddens 1984:284-6).

First, the interpretative or *hermeneutic mode of analysis* warrants renewed attention. This has to do with what Giddens calls the 'double hermeneutic'; social scientists have to deal with a reality which is already meaningful to or being interpreted by human agents who continuously reconstitute this reality. The term refers to the double process of interpretation which takes place when social scientists interpret the reality which has already been interpreted by the lay sociologist or geographer acting as human agent. This also implies that in practice it will be almost impossible to draw a sharp borderline between scientific knowledge and common sense knowledge. Consequently this means also that sources other than data, which are gathered in surveys or from statistical bureaux, are of use and important to social scientists. This accords with the distinction mentioned earlier between discursive and practical conscience and knowledge. Survey data could be designated as discursive knowledge in a propositional format. To understand discursive knowledge fully, however, one should also concentrate on other types of expression, such as jokes, humour, and sarcasm. In addition, non-verbal behaviour in all its variety should be taken into account because it is often a vehicle for practical knowledge

(Munters *et al.* 1985:50-51). It should be added that, for both kinds of knowledge, literary sources, cultural expressions, and the like are also indispensable. The methods of participant behaviour and very careful and close (qualitative) observation methods should become normal research procedures (for a good example see Mooij's [1987] Ph.D. dissertation).

Second, a new emphasis is being placed upon the *human agent*. This applies irrespectively of the focus that is selected to study reality: the individual human agent or the institutional level. The duality of structure in the context of the agent-structure-system interaction means, as we have seen, that the primacy of agent or system (institution) cannot be specified a priori. This creates a dilemma for the researcher: in the context of structuration theory it is not only theoretically impossible to separate agent and system but also inadmissible to accord them unequal importance in the production and reproduction of social reality in practice. Therefore the primary attention in research should go to the interaction of agent and system (Moos and Dear 1986:242). In research practice, however, it is almost impossible to avoid separation of system and agent. Giddens introduces the notion of 'bracketing', a methodological device which means that attention may be focused on either agent or system, provided that neither is considered superior and interaction between the two is the ultimate concern. Storper (1985), in a very readable and critical article about structuration theory, is rather sceptical about methodological bracketing but, as a device or rule of thumb, I cannot find fundamental fault with it. The analysis of agency means that we try to understand how individual actors draw upon structural elements in their interaction and thereby reproduce these structural elements (rules and resources). This requires that we become 'sensitive to the complex skills which actors have in coordinating the context of their day-to-day behaviour' (Giddens 1984:285). Dear and Moos (1986:356-8) introduce a very useful concept: *agency space*. Any particular event is the core of an agency space in which all agents having anything to do with this particular event take their positions. Two aspects of this agency space are important: the time dimension and the position of individual agents with respect to each other. Certainly when institutions are involved, decisions about a particular action are generally taken in several rounds which are separated in time. The time dimension is also needed to deal with decision-making processes that stretch in time and during which the focus of interaction may change. The position of agents with respect to each other is significant when we look at discrete events and try to understand how influential agents, ordinary agents, politicians, and so on play their respective roles. Both aspects tied together connect the *durée* of discrete events with the *longue durée* of institutional decision making and provide a theoretical apparatus for analysing the changing position of individual agents with respect to each other in the course of time. Furthermore, this way of looking at things makes it possible to incorporate institutions in the analysis of agency. Agents often operate as representatives of an institution; as such, they express not only their own views but also those of the institution that they represent. Within this framework, a series of events may be studied. The way in which results of a particular action become conditions for the next action (also unintended consequences and unacknowledged conditions) may be appreciated. And the boundaries of the agents' capability and knowledgeability may be brought to light. The notion of agency space offers the possibility of integrating institutions into the analysis of agency.

The analysis of institutions must deal with the role that institutions play in the reproduction of the *longue durée* and the relationship between institutions and individuals (Dear and Moos 1986:355). As already indicated, institutions act through individual agents who represent them. In fact it is impossible to analyse institutional behaviour without taking into account the agents who are associated with them. (An example is the study by Schat and Groenendijk [1982] on a dispute between a supermarket chain and two Dutch municipalities). On the other hand, the outcome of action at the institutional level is connected with the *longue durée*.

> The *longue durée* both provides the preconditions for understanding institutional action and is the setting through which institutional outcomes become manifest.
> (Dear and Moos 1986:355)

Being embedded in the *longue durée* also explains why institutional action has special significance for individual members of society. Institutions and their actions are an important part of the context in which human agents act.

After agency and institutions, the analysis of structural properties deserves mention. Structural properties of the social system are connected with the longest time perspective and may also stretch considerably over space. These properties are important with respect to the balance between enabling and constraining human actions. This leads to Giddens's third point. After the renewed attention for interpretive analysis and the new emphasis on the human agent, Giddens stresses the time-space constitution of social life and adds to this a plea for a disciplinary convergence (Giddens 1984:286). This is not treated further here because the previous paragraphs have elucidated the importance of time and space as constituting elements in the production and reproduction of society.

Summarizing the previous observations, there are three levels of analysis: structures, institutions, and agents (see also Dear 1987). Structures are the relatively constant and deeply embedded social practices which govern daily life (labour and capital relations, gender relations, the state). Institutions are the expressions of structures in reality which are stretched in space and time (state apparatus, multinationals, trade unions, local government, the family). Agents, finally, are the human actors who, by their actions, shape the outcome of social processes.

The position of regional geography: levels of analysis and the relevance of regions

In the context of structuration theory, space should be considered in terms of its involvement in the constitution of systems of interaction (Giddens 1984:368). In connection with the previous discussion, this view leads to a convergence of methodology and theory in the social sciences. The differences between geography, history, and sociology are now of another nature than previously. This was already implicit in the writings of Van Paassen, and, since Giddens, it has become explicit. The current position of geography could be evaluated in this light but, because the influence of theoretical developments is greatest in regard to the revival of regional geography, I prefer to focus on regional geography. Nevertheless, the points raised also refer to geography in general. After all, in terms of structuration theory, the task of regional geography touches the core of human geography.

As indicated, social reality may be analyzed at the level of structures, institutions, or agents. However, when space constitutes orientation towards social reality, the options should be extended to include the spatial scale at which processes operate. Dear (1987) proposes a distinction between micro, meso, and macro scales. This may be sufficient for heuristic reasons, but it is not satisfactory as a portrayal of reality. In fact virtually every process operates at its own spatial scale. Structure-institution-agent relations are found in locales at many different scales. Moreover, processes have different time perspectives both for their operation and in their impact. The situation becomes very complex because interactions and interdependencies exist between the spatial, temporal, and societal scales.

> Any narrative about landscapes, regions or locales is necessarily an account of the reciprocal relationship between relatively long-term structural forces and the shorter-term routine practices of individual human agents.
> (Dear 1987:12)

It should be added that the spatial expressions of an evolving society comprise one of these structural forces; social interactions are enabled, constrained, and constituted through space. According to Dear, 'the central problem in human geography is to understand the simultaneity of time and space in structuring social processes' (Dear 1987:15); in a footnote he adds that 'planning is the attempt to forge new time-space configurations in human affairs'. From this perspective, regional geography is not only at the heart of the geographical discipline, but the traditional distinction between regional and thematic geography becomes irrelevant. This does not imply that geographers should not specialize in economic systems, power relations, or human interaction and the ways in which these work and develop over space and time. It does entail concentrating on one of the key processes which are crucial to an understanding of the significance of space in the process of the constitution of society (see also Dear 1987:18-19).

Which regions are relevant to regional geography and how should they be delimited? These questions are regularly posed (for example in Buursink 1987). The questions may be tackled in two ways: either by delimiting relevant regions a priori, or by deciding about them a posteriori in relation to a process being studied. The problem is that any spatial process is scale-dependent in the sense that it has its own scale of operation, irrespective of any a priori fixed units. Thereby it is clear that the scale at which a process operates may change with time. This presents a problem for research. On the one hand, we know that our zoning systems should match the processes that we study. This entails developing a process-dependent research design and collecting data according to the protocol of that design for every single research project. On the other hand, to pursue longitudinal and comparative research (and of course for reasons of efficiency), we need a continuous registration system for data at different scales. This implies a choice for a priori given units. In the current situation we rely heavily on data that is published by national statistical bureaux and other (governmental) institutions. They regularly publish data using several zoning systems which often coincide with public administrative divisions. These spatial divisions are almost never contrived with our research aim in mind. The units in the zoning system are rarely homogeneous with respect to variables which are central to our research problem. And the units tend to be outdated with respect to the processes that we study. The investigator should take these limitations into account and try to use them to his advantage.

There is, however, another side to this problem. Some locales or regions are more important than others. Or, as we have seen earlier, some regions or locales are dominant in the sense that their constraining influence on human agency is very strong. This dominance is always related to the pervasiveness of a class of structural properties within the locale or region. A municipality is the spatial extension of local government; the Western world is the spatial extension of capitalism; and a state is the spatial extension of a certain style of government. These structural properties are a fundamental part of the context within which human and some institutional action takes place. Because these properties shape human action to a large extent, it is useful to demarcate such spatial divisions a priori and to collect statistics for them at regular intervals. The same applies to spatial divisions which can be made on cultural grounds, regions in which the inhabitants feel themselves very much at home (also emotionally) because of language, traditions, common history, and so on. A shared feeling of identity may also be dominant in the sense that it can influence human agency to a certain extent. These two kinds of region, for convenience called administrative and cultural regions, are very much part of our daily lives. That is precisely the reason why it is both practical and scientifically important to treat them as relevant regions in regional geography. Their delimitation, however, is difficult because the time-space stretching of the structural properties that identify them is not constant — their extensions in time and space are changing. This may have to do with the fact that human agents and institutions tend to establish and defend certain spaces as expressions of a way of life.

The two sides of the problem are hard to reconcile. A priori zoning systems are too inert and seldom correspond to the processes that we want to study. But some zoning systems can be determined a priori on the basis of their domination of a certain class of processes. On the other hand, it is far too time consuming and inefficient to delimit regions when needed (for research, planning, or other purposes), taking into account the time-space extension of the process under study or the event to be explained, although it is probably methodologically preferable to do so. Because of these considerations a myriad of zoning systems has been designed with rather specific purposes in mind. Data are gathered more or less regularly on such systems; these include nodal regions, such as metropolitan areas or commuter areas, and zonal regions, such as economic divisions or agrarian regions.

These types of regions are almost identical to those described by Buursink (1987:205-7). However, he sees the zonal and nodal types as scientific constructs and the administrative and cultural types as emerging from society. This distinction is acceptable because the two kinds of types have a different function in the daily life of the human agent. On the other hand, the zonal and nodal types are relevant to geographers only if they are closely aligned with the actions of the human agent (discursive or routine). It should be mentioned that physical geographical regions (delimited on the basis of climate, relief, or soils) may also be seen as dominant locales; the combination of 'structural physical properties' has a definite enabling and constraining influence on human agency.

Regional geography and, more generally, human geography have become more human and more contextual in the wake of structuration theory. Context is going to be one of the key concepts in geographical research. 'Context connects the most intimate and detailed components of interaction to much broader properties of the institutionalization of social life' (Giddens 1984:119).

This chapter has addressed aspects of the contextuality of social life and institutions. Actions in a time-space setting are surrounded by layers of contexts, and the discovery of their interdependencies is a challenge. This challenge should be taken up by developing a methodology in which the different levels of analysis (structures, institutions, agents), the different time perspectives, and the different spatial scales may be handled simultaneously. This requires a carefully balanced research programme in which small-scale comparative regional studies, which are aimed at the explanation of variety, are combined with more global contextual analyses. In the process of working toward these goals, the theory of structuration offers a fresh view of social reality and stimulates thought about human geography.

References

Buursink, J. (1987) 'Regionale Geografie: Nieuw of Opnieuw? *Geografisch Tijdschrift*, 21:198-212.

Dear, M.J. (1987) 'Reconstructing Human Geography', Paper presented at the colloquium on new aspects of social theory, Canadian Cultural Centre, Paris, March 1987.

— and Moos, A.I. (1986) 'Structuration Theory in Urban Analysis: 2, Empirical Application', *Environment and Planning A* 18:351-73.

Giddens, A. (1984) *The Constitution of Society, Outline of the Theory of Structuration*, Cambridge: Polity Press.

Gregory, D. (1984) 'Space, Time, and Politics in Social Theory: An Interview with Anthony Giddens', *Environment and Planning, D: Society and Space* 2:123-32.

— and J. Urry (eds) (1985) *Social Relations and Spatial Structures*, London: Macmillan.

Moos, A.I. and Dear, M.J. (1986) 'Structuration Theory in Urban Analysis: 1, Theoretical Exegesis', *Environment and Planning, A* 18:231-52.

Mooij, T. (1987) 'Interactional Multi-level Investigation into Pupil Behaviour, Achievement, Competence, and Orientation in Educational Situations', Ph.D. Dissertation, University of Nijmegen, Instituut voor onderzoek van het onderwijs, 's-Gravenhage.

Munters, Q.J., Meijer, E., Mommaas, H., van der Poel, H., Rosendal, R., and Spaargaren, G. (1985) 'Anthony Giddens, een Kennismaking met de Structuratietheorie', *Mededelingen van de Vakgroepen Sociologie* 14, Landbouwhogeschool, Wageningen.

Paassen, C. van (1976) 'Human Geography in Terms of Existential Anthropology', *Tijdschrift voor Economische en Sociale Geografie* 67:324-41.

— (1981) 'The Philosophy of Geography: From Vidal to Hägerstrand', in Space and Time in Geography: Essays Dedicated to Torsten Hägerstrand, *Lund Studies in Geography* 48, Lund: C.W.K. Gleerup.

— (1982) *Het Begin van 75 Jaar Sociale Geografie in Nederland*, Amsterdam: Sociaal-geografisch Instituut, University of Amsterdam.

Peeters, H.F.M. and Monks, F.J. (eds) (1986) *De Menselijke Levensloop in Historisch Perspectief*, Assen: Van Gorcum.

Schat, P.A. and Groenendijk, J.G. (1982) *Macht en Ruimte*, Haarlem: Romen.

Spaargaren, G., van der Poel, H. and Munters, Q.J. (1986) 'Het Oeuvre van Anthony Giddens: Centrale Thema's en Hoofdlijnen', *Sociologische Gids* 33:302-30.

Storper, M. (1985) 'The spatial and temporal constitution of social action, a critical reading of Anthony Giddens', *Environment and Planning, D: Society and Space*, 3:407-24.

Thrift, N.J. (1983) 'On the Determination of Social Action in Space and Time', *Environment and Planning, D: Society and Space*, 1:23-5.

7 Regional geography: between scientific theory, ideology, and practice, (*or*, What use is regional geography?)

Roger Lee

The current debate on regional geography is both more outward-looking and more restricted than that which surrounds Richard Hartshorne's (1939) advocacy of the central importance of chorography for geography. Although the chorography debate is extensive in alluding to fundamental distinctions in knowledge, it is inward-looking in applying these distinctions to the nature of geography as a discipline. By contrast, the contemporary debate is concerned less with the characteristics of knowledge and more with the nature of the relationships between space and society. It is concerned with trying to make sense of the constitution of a world in which geographical fragmentation and diversity are not only of demonstrable significance but also essential characteristics.

It is important to establish just what is meant here. It is all too easy to slip from a consideration of the local or the regional, the concrete and the specific within the constitution and dynamic of social reality, to the assumption that somehow the local regional is - or has become - more significant than other scales of analysis; that local/regional effectivity is paramount. This chapter tries to argue against such slippage. It suggests that the concern for the regional is entirely appropriate if the objective of social theory is to understand the contemporary world but that the trick is to place this concern within the wider context of social affairs.

Nevertheless, the increasingly widespread and insistent concern to construct more direct links between theory and the diversity of the empirical suggests a more utilitarian and firmly grounded agenda for social enquiry.[1] So, in discussing 'whither regional geography?', there is a clear sense in which we are discussing the usefulness of regional geography. A regional geography which is capable of demonstrating the effectivity of place in social process intensifies the significance of geography in social explanation and assists in the development of social theory that is sensitive to the inherent significance of space in social affairs (Cochrane 1987; Cooke 1987; Smith 1987b). The alternative title of this paper reflects this consideration. And there is another, more concrete (!) reason for this alternative. At the time of writing, Neil Smith's (1987a) account of the demise of geography at Harvard chimed both with the plea by Gordon Clark (1987) to personalize academic discourse and with the personal experience of distortion that is produced by the commodification of higher education and geographical research whereby the allocation of resources comes to reflect financial rather than academic excellence. Under the circumstances it is surprising that a concern for exchange value rather than use value does not inform the writing of this chapter!

In outline, the chapter argues that (regional) geography is not the missing link in social theory which finally makes sense of and justifies the geographical project (for geographers and non-geographers alike).[2] Accepting the social significance of space should not necessarily imply the elevation of regional geography to a position of primacy in social science. This is despite the fact that regional geography certainly enables an understanding of the relations between space and society and, in the process, sensitizes social and material thought to the kaleidoscopic nuances of cultural distinction and dynamics. The argument which is presented here is that the relationship between regional geography, theory, ideology, and practice is rather more down to earth. Regional geography insists upon the reality of a messy and fragmented world of complex determinations. In John Agnew's words:

> rather than generalizing from universal propositions about social process to a universal social form or social response, regional geography can provide a frame of reference for examining the relationship between causes and outcomes without the presumption of universality in outcomes.
>
> (Agnew 1987:xvi)

But regional geography also points to the holistic nature and material dimensions of social life, its making and constant transformation, and, in so doing, demonstrates the emancipatory potential of alternatives.

The primacy of (regional) geography?

Perhaps the most dramatic claim for the social primacy of geography is that made by Edward Soja:

To be alive is to participate in the social production of space, to shape and be shaped by a constantly evolving spatiality which constitutes and concretizes social action and relationships.[3]

(Soja 1985:90)

The surprise that is generated by this statement comes not from its content - certainly not that, of which more in a moment - but from the direct and central relationship that is implied between life itself and geography. A moment's thought reveals that the words mean everything and nothing: the statement is a monumental truism. To be alive is also to participate in the social production of 'nature' and time but to say so tells us little, if anything. What Soja appears to be saying, however, is slightly different and a little more meaningful, if hardly a revelation. It is, simply, that society can exist and function only if it *takes place*. In consequence, geography is a fundamental component of society. Soja himself expresses this notion rather differently and his materialist perspective shades his wording:

The realization that *social life is materially constituted in its spatiality* is the theoretical keystone for the contemporary interpretation of spatiality. [Emphasis in original]
(Soja 1985:94)

But he goes further to assert

a theory of social action activated by spatiality by ... spatial praxis

in which spatial praxis is defined as

the active and informed attempt by spatially-conscious social actors to reconstitute the embracing spatiality of social life.
(Soja 1985:116, 114-15)

Here we are back with the centrality of geography to life and the universe. Allan Pred (1984:287) goes almost as far in suggesting that the formation of individual biographies, the transformation of nature, and the establishment, reproduction, and transformation of power relations are inseparable from the 'becoming of place' as an 'historically contingent process'. What are we to make of such claims?

One attempt to make sense of them retreats from Soja's

ontological fundamentalism towards the concerns of geography as a discipline. Doreen Massey argues that human geography is concerned with three sets of relationships (social/spatial, social/natural, social/place — involving the coming together of society, space, and nature). She (1984a:1) suggests that the effectivity of these relationships is 'of great importance, not just for "human geography" but for the social sciences as a whole, and for what those social sciences are about — understanding, and changing, society'. The justifications that are offered for this claim are that:

(a) The spatial form of specific social processes and relation-ships affects the working of those relationships and processes as well as being shaped by them;
(b) Neither society nor nature may be conceptualized in isolation from each other;
(c) The local informs the general at the same time as the general affects the local. For example, 'regional specificity has an impact on the operation of general, national or international level processes' and 'the whole mosaic of regional specificities ... can have an enormous impact on the way that society "as a whole" ... is reproduced and changed' (Massey 1984a:9).

In contrast, 'the old regional geography', in its concern with uniqueness and specificity, posed the right questions but gave the wrong answers: 'Too often it degenerated into an essen-tially descriptive and untheorized collection of facts' (Massey 1984a:10, 11, 2). As a result, the significance of the notions that society is part of a world in which both space and nature are fundamental components was left unexplored. Geography matters not only because the discipline recognizes the diversity of the world but also because the geography of diversity is 'implicated in' (Gregory 1978:172) the very making of history and geography.

Now these assertions are highly seductive, especially to the geographer who is besieged with feelings of disciplinary inadequacy. However, the nature of the influence — precisely how (regional) geography matters to the making and remaking of history and geography — remains uncertain. Can we say that geography matters in the same sense that food and shelter matter, or production and consumption, or culture, race, religion, gender ...? Unless we are seduced by an extreme form of geographical fetishism, the answer to all of these questions must be 'no'. So how does geography matter?

Geography as life support?

Geographers have long argued that the incorporation of space into aspatial social theory undermines many — if not all — of the central assumptions of that theory. This view has been expressed most recently in a particularly dramatic way by David Harvey:

> The insertion of concepts of space and space relations, of place, locale, and milieu into any of the various supposedly powerful but spaceless social and theoretic formulations has the awkward habit of paralysing that theory's central propositions.

This condition is so debilitating that:

> Whenever social theorists actively interrogate the meaning of geographical and spatial categories, either they are forced to so many *ad hoc* adjustments that their theory splinters into incoherency or they are forced to rework very basic propositions.
>
> (Harvey 1985a:xiii)

The implication here is that geography matters, in that it shatters any attempt to understand society in geographical terms. All social theory must be geographical theory because geography shapes human affairs. It is impossible to conceive of social processes which are not socio-geographic processes. So historical materialism, for example, must become historical-geographical materialism.

But this conclusion and what might follow from it are not fully spelled out — at least in my reading of Harvey — and he devotes more attention to the (related) question of the relationship between theory and experience. In short, Harvey, when writing about the practice of social life, reveals an uncertainty about the significance of geography, which is obscured in his more conceptual statements by an apparent confidence in its central importance. Nevertheless, Harvey does suggest that it is not enough merely to consider capitalism *in* space. The questions of the production of space — 'how capitalism creates a physical landscape ... in its own image'[4] which then affects the dynamics of the circulation of capital — and 'the implications for political consciousness of such processes' form the more ambitious agenda that is explored in his two-volume 'historical geography of capitalism' (Harvey 1985a:xvii, xviii; 1985c).

The notion of the production of space is also taken up by Neil Smith (1986). Space is endowed with significance not because of any intrinsic ontological property but, he argues, because it is the product of socio-economic processes. It is not a 'dead "factor" but comes alive neither as a separate thing, field nor container but as an integral creation of the material relations of society' (Smith 1986:92).[5] Now, in an argument designed specifically to rebut 'Space blind reductionism' (p. 94), this is a curious claim. It puts space in its place, integral to but dependent upon material relations. Nevertheless, 'the role of geographical space' (p.88) remains important. Smith suggests that the 'mix of geographically expanded equalization and sharpened differentiation' during the world economic crisis 'represents the real solution adopted by capital' to the dilemma faced by a capitalism with no more pre-capitalist societies to penetrate. So Smith's space, like Harvey's, is active in presenting alternatives and (temporary) solutions to capitalist crisis. He argues that the 'transformation in the definition of developed and underdeveloped areas' and the 'hardened geography of crisis' (Smith 1986:100), involving the retreat of capital to the protection of strong home states, places '[G]eographical space ... more firmly at the centre of the economic and political agenda than at any point in recent history' (p.88). This 'role' for geographical space is similar to that argued at the regional level by Doreen Massey (1984b: ch.3), whereby capital may use locational decisions in a context of geographical differentiation as an integral part of, or alternative strategy for, production.

Despite Smith's claims for space, he is inconsistent and seems to fall into the trap for which he castigates others in trying to establish a separate sphere for space. Going on to argue that it is important to 'understand the limits to ... the renewed significance of space', he suggests that these limits are set by the even greater 'political significance' of the geographical restructuring of the world economy which 'overshadows' its 'spatial dimension' (Smith 1986:101). This too is curious: the 'political' issues that he mentions relate directly to conflicts over the spheres of influence of the Great Powers and their safeguarding of access to markets and resources for productive enterprises within their jurisdiction. Such conflicts arise directly out of the 'geography of uneven development' (p.88). Distinguishing the spatial and the political in this way diminishes Smith's claims for 'the role of geographical space' and helps to push it to the margins of the economic and political agenda. His distinction would not be accepted so readily by others who argue that geography matters and who stress the connections between the spatial and the social. Certainly, the stress upon the limits to space reveals a further uncertainty about the significance of geography in social affairs.

Is it generalizing too crudely to say that these formulations fall into the category of geography as additive — extending rather than reconstructing social theory? Maybe such a geographical reconstruction is not possible? Certainly it poses what for David Harvey (1985b) is 'the fundamental question to be resolved' in theorizing the historical geography of capitalism. He poses the question thus:

> Is it possible to construct a theory of the concrete and the particular in the context of the universal and abstract determinations of Marx's theory of capitalist accumulation?
>
> (Harvey 1985b:144)

Harvey's choice of 'urban' as a framework for analysis is particularly instructive here. Despite posing the question of choice, he does not answer his own question but instead defines what he means by urbanization — or rather the study of urbanization:

> The study of urbanization ... is concerned with processes of capital circulation; the shifting flows of labour power, commodities and money capital; the spatial organization of production and the transformation of space relations; movements of information and geopolitical conflicts between territorially-based class alliances; and so on.
>
> (Harvey 1985a:xviii)

The emphasis is clearly upon the primacy of the contradictory circuit of capital as a process rather than upon a reification of the 'urban' and there is now no mention of consciousness. The implication is that the geography matters, in that it extends ageographic theory.

These points may be exemplified by Harvey's discussion of urban politics which, symbolically enough, appears in *The Urbanization of Capital* rather than in *Consciousness and the Urban Experience*. Here Harvey is anxious to demonstrate that

> the urban community and its distinctive politics are produced under capitalist relations of production and consumption as these operate in and on geographical space.
>
> (Harvey 1985c:162)

To do so, he argues that urban politics has its 'place' in 'the geography of uneven capitalist development' (ch.6). If the

> fundamental Marxist conception ... is of individuals and social groups ... perpetually struggling to control and enhance the historical and geographical conditions of their own existence

then putting the geography into the analysis of struggle

> immediately triggers concern for the urbanization of capital as one of the key *conditions* under which struggles occur. It also focuses our attention on how capitalism creates spatial organization as one of the *preconditions* of its own perpetuation. [My emphasis]
> (Harvey 1985c:163-4)

Here we see the assertion of the primacy of the economic over both the political and the spatial. In Harvey's view, space provides 'conditions' and 'preconditions' even though he claims that, far from 'disrupting the Marxian vision, the injection of real geographic concerns enriches it beyond all measure' (p.164). Again, however, it is difficult to escape the conclusion that, despite his initial assertions about the paralysis of ageographic social theory and the panegyric on the value of geography to Marxist theory, geography is conceived as an extra ingredient — improving but not rewriting established ageographic theory. It provides life support but not life itself.

Paradoxically perhaps, the same comment may also apply to the so-called locality studies in which the emphasis is laid upon the working out of broad social processes as they manifest themselves at the local level:

> Research in localities enables the finer-grained impact of local economic restructuring to be clearly understood.
> (Cooke 1986:2)

As Neil Smith puts it:

> The point is ... to construct a multidimensional portrait of the spectrum of change in the entire social fabric.
> (Smith 1987b:61)

To be sure, uniqueness is not only recognized but also actively sought out:

> Localities are distinctive sociological entities and ...
> although they are increasingly subject to structural change,
> this does not mean that their social and cultural character-
> istics can be reduced to the causes of such change.
>
> (Urry 1986:2)

But it is a uniqueness which is subject to sociological determinism and which tends to privilege pre-existing concepts,
especially those of class, above the context of the localities
themselves. As a result, not only do the localities become
dependent but also much of their richness, significance, and
meanings as localities is lost (Rose 1987; see also Savage 1987).
And yet this is not to accept, with Smith (1987b:62), that 'the
project is primarily about the localities in and of themselves'.
Gillian Rose (1988) argues for the need to recognize that the
lines of social causation are neither single track nor straightforward and that the study of localities helps to elucidate the
diversity of social determination. She is able to show that the
radical nature of politics within the inter-war London borough
of Poplar cannot be explained simply by reference to working-
class political consciousness. Indeed, the role of class-based
organizations such as trades unions was contradictory in this
respect. But the local intersection of a variety of influences
(religion, citizenship, feminism, and pacifism, as well as class)
combined in a particular way for a time in this place to
generate the distinctive politics of locality. And the point is
that locality itself was highly influential in the way in which
this combination worked out in practice.

Thus the sensitivity to local circumstance which is enabled
by locality studies does begin to reveal the complexity of the
relations between concepts such as class and local civil society.
An example is provided in the work of Jane Mark-Lawson,
Mike Savage, and Alan Warde (1985). They are able to reveal
the complex and subtly variable ways in which gender and
class interact to markedly different political effect within
different localities, each structured and differentiated
primarily by the nature of production relations.

Alan Warde (1986) has generalized this theme to suggest
that the nature of the relationship between politics and civil
society may be elucidated by local processes of political
communication which draw upon the complex of cultural and
political practice within the locality as well as that of
economic practice. And this nation resonates with wider
questions concerning regions and the conditions of existence
of social practice. It begins to suggest, perhaps, that localities
and regions may provide life itself and not just life support.

Geography as life itself?

So there may be a sense in which geography matters even more. Harvey's discussion of the conditions of existence of social struggle recalls the use by Cutler *et al.* of the same concept in discussing the nature of social formations:

> The social formation is not a totality governed by an organizing principle, determination in the last instance, structural, causality, or whatever. It should be conceived as consisting of a definite set of relations of production together with the economic, political, and cultural forms in which their conditions of existence are secured. But there is no necessity for those conditions of existence to be secured and no necessary structure of the social formation in which those relations and forms must be combined.
>
> (Cutler et al. 1977:222)

Here the emphasis is not merely upon the primacy of social relations in social analysis (of which more later) but upon their contingent dependence on conditions of existence within a social formation. Whether or not these conditions of existence are secured is an empirical rather than a theoretical or logical question. It is dependent upon social struggle which has its own geography (e.g. Hobsbawm and Rude 1973; Charlesworth 1979) as, indeed, do the conditions of existence themselves (and Cutler *et al.* miss out one of the most important, namely nature).

This argument suggests that, to use Harvey's example, it is not only struggle (as one of the consequences of an established set of social relations) which is affected by the geographical conditions under which it occurs, but whether or not a particular set of social relations may ever be established and find expression at a particular place and time. Here it is arguable that the geographic is of prime importance in providing or not providing those conditions of becoming.[6] So Allan Pred asks, 'What is the common place?' and answers his question thus:

> It is a process whereby the reproduction of social and cultural forms, the formation of biographies, and the transformation of nature ceaselessly become one another, at the same time that time-space specific path-project intersections and power relations continuously become one another.
>
> (Pred 1984:292)

The distinguishing characteristic of Anthony Giddens's structuration theory is, according to Derek Gregory (1986:467), that he 'accords a central place to *time-space relations* in social theory'.[7] This arises from his insistence upon the duality of system and structure — whereby structure is involved in the production of action — rather than upon dualism in the relationships between structure and agency. Structures become as the product of human action but humans draw upon structure in the practice of their daily lives and, it might be added, in the conscious attempt to change the relationship between structure and agency. Regional geography is crucial in this project. In Nigel Thrift's terms, the region may be regarded as the

> 'actively passive' ... meeting place of social structure and human agency, substantive enough to be the generator and conductor of structure, but still intimate enough to ensure that the 'creature-like aspects' ... of human beings are not lost.
>
> (Thrift 1983:38)

In similar vein, Torsten Hägerstrand refers to

> the world as a fine-grained *configuration of meeting places* rather than as a system of regions or of aggregate categorical variables.
>
> (Hägerstrand 1984:378)

Action is nothing unless it is concrete: it must take place. It must, therefore, be situated in, draw upon, and thereby modify the setting for social interaction. Giddens (1984) refers to this setting as the locale.

However, if we take the region seriously, it may in practice short-circuit structure and crush the 'creature-like' aspects of human beings. In any event, the region provides the setting for struggle to make history and geography: to shape social relations and to make them effective. *The Fatal Shore* (Hughes 1987) was quite literally that for Aborigines and transportees alike as they contested the meaning of Australia. Their struggles to create new meanings or to reassert the old were made necessary and were informed by their mutually incomprehensible 'geographical unconscious' (ch.3). This was not reducible only to their own limited geographical experience or imagination, not merely the product of their own geographies and histories which had to be unlearned or asserted in the new region. It was and is an actively contested unconscious, to be used and modified in establishing conditions for existence in the dynamic and conflict-ridden process of social interaction. Torsten Hägerstrand expresses something of the same when he writes:

> the knowledge we have is continually used for bringing about or preventing changes in the world. ... Every action is situated in space and time and for its immediate outcome is dependent on what is present and absent as help or hindrance when the events take place. The secondary consequences in their turn are dependent on a new set of presences and absences and so it goes further and further out from the initial action.
>
> (Hägerstrand 1984:377)

And if Doreen Massey's (1984b) 'layers of investment' are interpreted in terms of the duality of structure and agency in and across particular settings, there are resonances here too.

These comments stress the significance of the region for the strategic as well as the routine aspects of human inter-action and indicate that the contours of the relations of power are not given. At one and the same time they may be regionally, geographically, historically, and socially dependent, they must be contested, and so they may be changed. But the region is never merely given; it is always and everywhere a social product. As such it too is a contested terrain: a locus of struggles (not reducible simply to the economic) to endow meaning upon the region, to gain power over it and so, in either event, to transform it. The Australian example is relevant here but so too is a whole host of others, ranging from the global geopolitical scale (e.g. the new imperialism of the external relations of the EC) through the regional/urban (e.g. the role of localities in initiating and shaping the emergent welfare state in Britain before 1948 [Lee 1988b] to the local (e.g the struggle over the meaning of localities such as the docklands of east London; see, e.g., Ambrose 1986).

It is at this point that I have some difficulty with Nicky Gregson's (1986) critique of Giddens. Two points of disagree-ment, or rather misunderstanding, are helpful in the present context. Gregson argues that Hägerstrand's (1982) time geography is a contextual project whilst that of Giddens is a form of compositional theory.[8] For Hägerstrand, the region is given to experience whilst, for Giddens, the region becomes through experience and practice. Now it may be that Giddens is concerned with second-order issues (those relating to the nature of human society) whilst Hägerstrand is expressing a first-order concern with more factual issues. But to accept uncritically the notion of social life 'as it is' in Hägerstrand's work (or, indeed, in anybody else's) denies that concepts are necessary even to express or experience what is and what is not present in Hägerstrand's diorama — the *'thereness'* (1982:326) of entities. The concepts that we bring to experience help to endow meaning even if we have to be 'educated' in the meanings that are 'acceptable' in our society. With these concepts, the social relations from which they spring and the material reality to which they give rise, we continuously produce and transform geographical space — a space that informs our actions and is shaped by them.

This, then, is a defensive rewriting of Giddens on the grounds that compositional theory and contextual theory form a duality and not a dualism. Both are necessarily interdependent in attempting to make sense of social practice: we make and remake regions at the same time as we are shaped and reshaped by them. Where, therefore, do we locate determination?

Gregson (1987b) suggests that the denial of 'real' structures (structures independent of social interaction) in structuration makes it impossible to identify causal and internally homogeneous entities for empirical analysis. As a result, she stresses what I would argue is another false dualism between structuration theory and realist concepts of necessary relations and rational abstractions. This dualism is false because social relations are both an *ontological* necessity for social life — a real structure inescapably given to us — and, in the practice of social life, a product of social interaction — we can and do constantly change them (Lee 1988a).

Furthermore, this process of transformation is highly creative. It is in the struggles to establish/maintain/control/change/transform social relations that we may find the locus of determination. So we can say with Henry Lefebvre, that 'each epoch produces its own space' (Burgel and Derzes 1987:31). At the same time, however, whilst social relations are necessary to social life, there is nothing inevitable about the emergence of a particular set of social relations and there is no simple one-to-one relationship between each epoch and its space. This is so not only because of the recursive duality of structure and agency but also because the relationship between them is of 'incredible complexity and subtlety'. Hence the current interest in the locally specific and the regional; in difference and 'reversal'. This interest is a product in part of the desire to overcome the stultifying effects of theoreticism, abstraction, structuralism, and generalism; in part of the need to confront the complexities of post-Fordist geographies; and in part of the wish to celebrate diversity and uniqueness — a diversity with profound macro-social and political implications.

It is Lefebvre's 'incredible complexity and subtlety' of the geography of structure and agency, rather than a dualism of structuration and necessary relations, which makes regional description and the assessment of regional significance so very difficult, vital, and, apparently, so tantalising to contemporary social theory. And it is this complexity and subtlety which suggests that a theory of social action cannot be *activated* by spatiality in the exclusive sense implied by Soja. Social life is far too complex and subtle to allow such exclusionary tactics.

So what use is regional geography?

Certainly, regional geography provides us with a reminder of the complexity of social life. Thus Allen Scott and Michael Storper advocate not only an economic geography more in tune with the complexities of the social world but also a reordering of social theory around the regional concept:

> As capitalism opens out from a broad general structure into a system of geographically differentiated units of organization it takes on the form of an extraordinarily varied mosaic of socio-spatial relationships. In turn the retotalization of capitalism in a way that incorporates this mosaic provides us with a dramatically widened conception of historical and social change.
>
> (Scott and Storper 1986:14)

Thus (regional) geographical analysis addresses an inherent feature of the social world and insists upon the significance of the geographical continuum of scale. Whether or not it is necessary and possibly even a distortion to invoke a specifically '*meso* level of theory and empirical enquiry', the insistence upon geography forces us to go beyond both top-down (structural) bias in social explanation and accounts which are built up from the aggregation of small-scale events. In short, it focuses attention upon the significance of social interaction which must be geographical and which takes place continuously up and down the continuum of geographical space.

Not only does regional enquiry address an inherent ontological feature of the social world, it also enables an appreciation of the significance of uneven development. The 'territorial production complexes' (Scott and Storper 1986:310), which are produced by this geographically-informed process of development, frame the processes of both urban and regional development and provide the material underpinning of the dynamics of international relations. Under these circumstances regional geography is not merely of analytical significance but of paramount emancipatory importance.

This leads to a final point. We live at a time when the capitalist world is passing through a crucial divide: from modernism to post-modernism; from the fourth to the fifth Kondratiev; from Fordism to neo-Fordism; through *The Second Industrial Divide* (Piore and Sabel 1984) when *The End of Organized Capitalism* (Lash and Urry 1987) is nigh; and from Fordist to flexible accumulation. Whatever the true form of the transition, it implies a much greater sensitivity and responsiveness to geography on the part of capital. Indeed it could be argued that the restructuring of the social relations of production and the spatial implications of the new technology place (regional) geography at the heart of the current transition in a way far less contingent than is suggested by Neil Smith. Certainly, not only is the geography of the current transformation remarkably varied and dynamic, but geographical restructuring is a centrally important feature of it. In accounting for the change from organized to disorganized capitalism, Scott Lash and John Urry (1987:84) argue that 'there is in the spatial an aspect of social relationships', that there are 'particular spatial patterns associated with *each* of these phases of capitalist development', and, in elucidating 'what it is like to be "modern", to live in a modern age', they stress the 'extraordinary spatial and temporal transformations ... that substantially organize our experience of modern social life'.

Not surprisingly, then, the diversity of transition and the fragmentation of previous wholes has led to an increased concern with the geographically distinctive and local and has encouraged a mode of theorizing which stresses fragmentation and unintended consequences. This, I would argue, is extremely dangerous. 'We still live in a world dominated by capitalism', and 'broad social change' cannot be reduced to the 'inchoate swirl of human agency' (Harvey and Scott 1989). Related to this is Neil Smith's (1987b) critique of realism in which the distinction between necessary and contingent relations leads to the conclusion that, as a product of contingent relations, geography is not accessible to abstract theory.

If our objective is to change rather than merely to understand the world, the significance of regional geography lies in reminding us of the ever-present possibilities of making alternative geographies. If the 'essence of the intellectual enterprise we are engaged in is to construct sustainable generalizations', regional geography enables us to 'judge when these generalizations are no longer sustainable' (Smith 1987b:67). And it reminds us of the new possibilities for challenging capital which are presented by its new realities — not least the potential that is released by computer-based telecommunications for the decentralization of knowledge, despite the probabilities of increased concentration. More than ever we must not — as Harvey and Scott (1988) seem to imply — simply remain spectators at 'the global waltz of capital's autonomous self-activating development' (Cleaver 1979:27). Participation, resistance, and alternatives should form the basis of our project and regional geography informs all three.

Notes

1   This is a concern both to avoid reductionist accounts and to engage the geographical complexity of the social world by escaping from the stifling grasp of grand theory. See, for example, Gregson (1987a; 1987b), Leitner (1987). Most famous perhaps is Castells' (1983:xv, vxi) wish to bridge the 'increasing gap between urban research and urban problems' by analysing 'the relationship between people and urbanisation'; but see Harvey and Scott (1988).
2   It is difficult to distinguish, from within the contemporary chorus of invocations for geography, the voices singing the virtues of regional geography from those exclaiming more generally that geography matters. Indeed, this chorale is, perhaps, more like a fugue, in that different groups appear only to be singing different parts which are finally resolved together. The distinction will therefore remain blurred in this paper, just as it is left by many writers in the field.
3   Soja (1985:123) defines spatiality thus: 'socially-produced space, the created forms and relations of a broadly-defined human geography'.
4   In the companion volume (Harvey 1985c:162) this concern is extended to the 'physical and social landscape'.
5   The point here is not to debate whether space is a material or non-material product but to emphasize the notion that space is socially produced.
6   However, I do not accept Stuart Corbridge's (1986) overly simplistic extension of such arguments which suggests that

the outcome of capitalist development is entirely contingent upon its conditions of existence. There is a logic to capital accumulation which, if subverted, prevents accumulation from taking place. If accumulation does take place then the logic must have been facilitated in some way or another.

7   Whatever the problems of eliding space and time in this way (see Lash and Urry's [1987:ch.4] discussion of this point), we shall ignore them here and instead concentrate upon the centrality of space and time in social action.

8   Contextual theory is an approach stressing the significance of time and space (or, to take Giddens's concept, time-space) for explanation whereas compositional theory uses prior categories.

References

Agnew, J. (1987) *The United States in the World Economy: A Regional Geography*, Cambridge: Cambridge University Press.

Ambrose, P. (1986) *Whatever Happened to Planning?* Harmondsworth: Penguin Books.

Burgel, G. and Derzes, M.G. (1987) 'An Interview with Henri Lefebvre', *Environment and Planning, D: Society and Space* 5(1):27-38.

Castells, M. (1983) *The City and the Grassroots*, London: Edward Arnold.

Charlesworth, A. (1979) *Social Protest in a Rural Society*, Historical Geography Research Series 1, Institution of British Geography, Historical Geography Research Group, Norwich: Geo Books.

Clark, G.L. (1987) 'On Being an Academic in America', *Environment and Planning, D: Society and Space*, 5(1):Editorial.

Cleaver, H. (1979) *Reading Capital Politically*, Brighton: Harvester Press.

Cochrane, A. (1987) 'What a Difference the Place Makes: The New Structuralism of Locality', *Antipode* 19(3):354-63.

Cooke, P. (1986) 'The Changing Urban and Regional System in the United Kingdom', *Regional Studies* 20(3):243-51.

— (1987) 'Clinical Inference and Geographic Theory', *Antipode* 19(1):69-78.

Corbridge, S. (1986) *Capitalist World Development*, London: Macmillan.

Cutler, A., Hindess, B., Hirst, P., and Hussain, A. (1977) *Marx's Capital and Capitalism Today*, London: Routledge & Kegan Paul.

Giddens, A. (1984) *The Constitution of Society*, Polity Press, Cambridge.

Gregory, D. (1978) *Ideology, Science and Human Geography*, Hutchinson, London.

— (1986) 'Structuration Theory', *Dictionary of Human Geography* Oxford: Basil Blackwell.

Gregson, N. (1986) 'On Duality and Dualism: The Case of Structuration and Time Geography', *Progress in Human Geography* 10(2):184-205.

— (1987a) 'Human Geography and Sociology: Common Ground or Common Object?' *Political Geography Quarterly*, 6(1):93-101.

— (1987b) 'Structuration Theory: Some Thoughts on the Possibilities for Empirical Research', *Environment and Planning D: Society and Space* 5(1):73-92.

Hägerstrand, T. (1982) 'Diorama, Path and Project', *Tijdschrift voor Economische en Sociale Geografie* 73(6) 99:323-39.

— (1984) 'Presence and Absence: A Look at Conceptual Choices and Bodily Necessities', *Regional Studies* 18(5) 373-79.

Hartshorne, R. (1939) *The Nature of Geography: A Critical Survey of Current Thought in the Light of the Past*, Lancaster Pa: Association of American Geographers.

Harvey, D. (1985a) *Consciousness and the Urban Experience*, Oxford: Basil Blackwell.

— (1985b) 'The Geopolitics of Capitalism', ch. 7 in D. Gregory and J. Urry (eds) *Social Relations and Spatial Structures*, London: Macmillan.

— (1985c) *The Urbanization of Capital,* Oxford: Basil Blackwell.

— and Scott, A.J. (1989) 'The Practice of Human Geography: Theory and Empirical Specificity in the Transition from Fordism to Flexible Accumulation', in W. MacMillan (ed.) *Remodelling Geography*, Oxford: Basil Blackwell.

Hobsbawm, E. and Rude, G. (1973) *Captain Swing*, Harmondsworth: Penguin Books.

Hughes, R. (1987) *The Fatal Shore*, London: Collins Harvill.

Lash, S. and Urry, J. (1987) *The End of Organized Capitalism*, Cambridge: Polity Press.

Lee, R. (1988a) 'Social Relations and the Geography of Material Life', in D. Gregory and R. Walford (eds) *Horizons in Human Geography*, London: Macmillan.

— (1988b) 'Uneven Zenith: Towards a Geography of the High Period of Municipal Medicine in England and Wales', *Journal of Historical Geography* 14(3).

Leitner, H. (1987) 'Urban Geography: Undercurrents of Change', *Progress in Human Geography* 11(1):134-46.

Mark-Lawson, J., Savage, M., and Warde, A. (1985) 'Gender and Local Politics', ch. 11 in Murgatroyd *et al.* (eds) *Localities, Class and Gender*, London: Pion.

Massey, D. (1984a) 'Introduction' in Massey, D. and Allen, J. (eds) *Geography Matters!*, Cambridge: Cambridge University Press.

— (1984b) *Spatial Divisions of Labour*, Houndmills: Macmillan.

Piore, M.J. and Sabel, C.F. (1984) *The Second Industrial Divide*, New York: Basic Books.

Pred, A. (1984) 'Place as Historically-Contingent Social Process: Structuration and the Time-Geography of Becoming Places', *Annals of the Association of American Geographers* 74(2):279-97.

Rose, G. (1987) 'Locality and Culture in Poplar', Paper to the conference on *Locality and Politics in Inter-War Britain*, Queen Mary College, London, April 1987.

— (1988) 'Locality, Politics and Culture: Poplar in the 1920s' *Environment and Planning, D: Society and Place* 6:151-68.

Savage, G. (1987) *The Dynamics of Working Class Politics: The Labour Movement in Preston 1880-1940*, Cambridge: Cambridge University Press.

Scott, A.J. and Storper, M. (1986) *Production, Work, Territory. The Geographical Anatomy of Contemporary Capitalism*, Winchester, Mass.: Allen & Unwin.

Smith, N. (1986) 'On the Necessity of Uneven Development', *International Journal of Urban and Regional Research* 10(1):87-104.

— (1987a) '"Academic War over the Field of Geography": The Elimination of Geography at Harvard, 1947-1951', *Annals of the Association of American Geographers* 77(2):155-72.

— (1987b) 'Dangers of the Empirical Turn: Some Comments on the CURS Initiative', *Antipode* 19(1):59-68.

Soja, E. (1985) 'The Spatiality of Social Life', ch. 6 in D. Gregory and J. Urry (eds) *Social Relations and Spatial Structures*, London: Macmillan.

Thrift, N. (1983) 'On the Determination of Social Action in Space and Time', *Environment and Planning D: Society and Space*, 1(1):23-57.

Urry, J. (1986) 'Locality Research: The Case of Lancaster', *Regional Studies* 20(3):233-42.

Warde, A. (1986) 'Space, Class and Voting in Britain', pp.33-61 in K. Hoggart and E. Kofman (eds) *Politics, Geography and Social Stratification*, London: Croom Helm.

8 The challenge for regional geography: some proposals for research frontiers

R.J. Johnston

To many people, the issue of how to do research in regional geography raises few methodological let alone epistemological problems. To them, the practice of regional geography falls squarely into the empiricist conception of science — though they may not recognize it as such. Regional geography involves bringing together factual material: its data collection involves a combination of straightforward observation — recording what is there — and using the results of others' observations (such as census reports); its methodology involves arranging the material in an organizing framework. In many cases, and this was certainly so in Britain in the heyday of regional geography, such a framework implied a particular theory (albeit poorly articulated) about how humans use the earth, because primacy in the presentation was given to the physical environment. A standard presentational sequence was widely used: geology, landforms, climate, soils, vegetation, population, etc. The subsequent failure to generate intellectual excitement led to the criticisms that were made by Freeman (1961), Gould (1979), and others (see Johnston 1984a).

Against this empiricist tradition, others have argued that regional geography involves methodologies that can be equated with what we now widely term humanistic geography. To them, regional geography is not just the collection, collation, and presentation of facts but an exercise in synthesis and interpretation. The role of the regional geographer is to transmit to others the characteristics of regions. To do this involves: identifying the salient elements that go to make up the regions; bringing these together in a synthesis that both allows the definition of regions and highlights their features; and preparing a narrative (Daniels 1985) which conveys the elements of those regions to the audience, whether it be students or the population at large. This is the sort of regional geography for which Hart (1982) calls. It essentially follows the canons of idealism, as outlined for geographers by Guelke

(1974). At its best, it can provide very readable, personal views of a place — supposedly more objective and therefore more scholarly than descriptions evoked by novelists, poets, and painters. But to me, however enjoyable the read, such work is almost always ultimately disappointing because its goal is insufficient: it can inform and entertain but can it advance knowledge?

Our goal is to promote regional geography — if we want to call it that — as a part of the educational process. Certainly we want to inform; no doubt we want to entertain; but much more than that, we want to understand. Hence my remarks here are directed towards that basic question: how can we use regional geography to promote understanding?

Purpose

The first task that I want to address, therefore, is to ask what we want of regional geography. I do not pose this in the narrowly utilitarian way in which questions are asked about what is done in so many areas of education today. Indeed, I would want to differentiate carefully between *training*, which is the acquisition of particular, applicable skills, and *education* which is the development of the individual. My concern here is entirely with the latter. I accept that training is an indispensable part of the preparation of people for life but I would claim that it is insufficient unless we see the role of the education system (broadly defined) solely as preparing individuals to occupy particular niches in an economic system. The sort of education for which I am arguing helps people to understand the society in which they live, as well as preparing them for survival in it. The latter is undoubtedly necessary but without the former we abuse ourselves and our society.

What role, then, can regional geography play in this educational process? Clearly it has a factual element to it, which I take for granted. One task of regional geography is simply to inform people about what is where in the world. Undoubtedly we do not do that very well at present, because so many people — from Presidents of the USA down — are woefully ignorant of basic geographical facts. But perhaps we do it so badly because of how we present the material. Mere portrayal of facts is insufficient to capture attention: those facts should be used. Hence I pass on to other tasks, and assume that learning the basic geographical facts is integral to them.

I am going to focus on two roles for regional geography.

These are set in a framework of why we need education, not just as an end in itself but also as a means of survival for the species. In one sense, therefore, my advocacy of regional geography is deeply utilitarian — but not in the usual way in which the term is used.

Distancing, places, ignorance, and conflict

In their seminal work on social areas in cities, Shevky and Bell (1955) introduced a concept of the *increasing scale of society*. By this they meant the growing complexity and density of life: individuals occupy very narrowly defined niches within society, are therefore necessarily in contact with many others, and live in very close proximity to them as a consequence. This is clearly a description of the process of urbanization, but it implies much more than just the congregation of people into high density agglomerations; it describes structural and behavioural as well as demographic trends. We are frequently told today that urbanization has come to an end, and that agglomerations are no longer necessary to the conduct of 'advanced, post-industrial societies'. That contention is dubious, I believe, and certainly is not yet confirmed empirically. To the extent that counterurbanization has taken hold, I wonder if it might be as a part of the processes that I am going to describe here?

How do people cope with such complexity? It is clear that they must have some strategy for simplifying and surviving in it, because the mass of information that is available cannot be handled in any other way. They cannot come to know all of the people with whom they necessarily come into contact, simply because the number exceeds their capacity to store all of the information about people and to recall the relevant portions at the relevant time. Nor can they expect that the people with whom they will next come into contact are those whom they have met and come to know previously: they must be able to survive in a world of strangers.

There are two aspects of the usual coping strategy that I want to stress. The first is *withdrawal into the known*. We organize our lives — and have our lives organized for us — in ways that are based on allocating ourselves to cells. Thus most large and certainly the most successful organizations are structured hierarchically, with each person in the hierarchy having between seven and ten people responsible to him or her. This is how many of the large ecclesiastical institutions operate; it is how most armies are run; and it is how most factories and offices are structured. Outside such organizations, the hierarchy may be absent (or at least be relatively suppressed) but the size of the cells remains small. Is it purely coincidental that the rise of the nuclear family as a (if not the) basic social unit parallels the increasing scale of society as represented by mass urbanization?

The second aspect that I want to stress is the *stereotyping of the unknown*. To the extent that we live in our small cells, so we survive among people with whom we are in regular and frequent contact and whom, therefore, it can be said that we know. Here I differentiate between knowledge and acquaintanceship because, in vernacular usage, to know a person frequently implies just being acquainted with her or him. For some people, that is sufficient and they rarely venture outside their cells. For the vast majority, however, frequent external contact is necessary, either with people with whom they are acquainted only or about whom they have no personal information at all. How do they handle situations where contact with a total stranger involves more than a fleeting transaction (such as buying a train ticket) and which need not be personal at all indeed, increasingly is not?

The commonest way of handling such situations is by the use of stereotypes: people are treated as members of a category which shares certain traits and characteristics. Such stereotypes are widely used in all aspects of life. Definition of an individual by one or more of a range of characteristics — racial, occupational, age, etc. — carries with it connotations of certain behavioural features: professors are absent-minded; young people like loud music; football fans are violent. Such connotations may be neutral, e.g. Swedes are blond, but they rarely are. Usually they carry either positive or negative interpretations. This is clearly exemplified by the treatment of Jews over many centuries; by the current treatment of blacks by Afrikaners in South Africa; by the representation of the Russians by the NATO powers as 'a threat' to world security; and by jokes about the Irish or other ethnic groups.

All of these stereotypes are social creations: they emphasize characteristics that are not innate. What concerns us, therefore, is how they are created and what their consequences are. My argument here is that the processes of withdrawal into the known and of stereotyping of the unknown are linked, and that unravelling the link is an important task for regional geography.

According to my presentation of the withdrawal process, life is organized in cells. (We are all members of several cells, of course, some of which overlap and others of which are mutually exclusive.) Those cells need not be, but very frequently are, spatially defined. As Bob Sack (1983; 1986) has so clearly argued, territoriality is an important strategy to be used for a whole range of social, economic, and political ends. People define themselves, i.e. make statements about who they are, by identifying themselves with places: they are what they are because of where they are. Others are defined in that way too: they are required to occupy certain territories, which are used to label them. Thus when we withdraw into the known, or are required so to do, very often that withdrawal is spatial: we withdraw into a particular place which is endowed with certain characteristics. In other words, a particular type of person lives there.

Having withdrawn spatially, our stereotyping then follows: the people outside our territory are those from whom we have retreated and whom we then characterize out of relative ignorance. Thus territoriality is both a strategy for organizing our life spatially and a way in which we develop means of dealing with others.

So far, I have presented these two strategies of withdrawal and stereotyping as if they were initiated *de novo* by every individual or generation. Clearly that is not the case, since the strategies are already in place. We are born into the consequences of such strategies, and we are socialized accordingly. We learn to be members of our cells — our regions, or places — and we learn images of others accordingly. Thus the strategies are self-perpetuating. This is not deterministic, however, because we can reject our cells and their images of others. But socialization in places is a powerful process, and the spatial (and hence social) confinement of how we learn to be people — a process that is continuous through life, but almost invariably slows down — is a major way of reproducing a divided society which is built on withdrawal and negative stereotypes of outsiders.

As yet, I have said nothing about *spatial scale* in this discussion, and it could be implied that much of what I am presenting is related to micro-scale behaviour only. But, as societies have become more complex, as interdependence has increased, so the spatial scale of societal organization has become wider and our local cells have been nested into larger cells. Thus we are now at one and the same time members of street, neighbourhood, city, county, regional, national, etc. communities, and each of these has its own territory and its own images of others. As regional geographers, we must focus on them all.

I want to stress one goal of regional geography that, to me, is highly utilitarian. An almost inevitable consequence of withdrawal and stereotyping is the misunderstanding and mistrust that follows ignorance: because we do not know people, we tend to attribute motives to them that are against our interests. Such misunderstanding and mistrust can readily lead to conflict, the ultimate consequence of which is violence. We see this best at the international scale but no doubt we can all identify local examples. Most of these will not have led to violence but, if not resolved, will have led to continued unease and unhappy relationships.

A major role of education must be to dispel the misunderstanding and mistrust. Education must pull people out of their cells. It cannot do this literally (except at certain scales, and geographical fieldwork is clearly one method) but it can do it in many other ways. Its goal must be to advance understanding both of others and of self. It must promote awareness, and the end of stereotypes. And what can be more vital? Given our power now to destroy not only ourselves but also the ability of our planet to support life, if awareness is not enhanced and misunderstandings are not removed, we may no longer have a geography to teach or any geographers to teach and be taught.

Uncovering how the world works

The case that I have made in the preceding section is for a geography that advances self- and mutual understanding, thereby contributing to the reduction of ignorance and mistrust in the world, and so helping to reduce conflict and to ensure the perpetuation of life on earth. To me, that is a deeply utilitarian programme; it is geography with a clearly applied purpose. But it is insufficient: we need to do that ... and more.

The general goal that I am promoting is an improvement in the human condition and in people's life chances. Advancing understanding in the ways outlined above should contribute to that. However, the potential contribution of geography is substantially constrained, not so much by the ability of geographers to perform the task but rather by the economic, social, and political systems within which they operate. If, as I believe to be the case, capitalist societies can survive only by promoting inequalities over a very wide range of elements of the human condition, and also by promoting increased despoliation of the environment, then the sources of conflict will remain. Our success at increasing understanding may help to mitigate current difficulties and to resolve potential crises but it will not deflect the main dynamic of capitalism. To achieve that, and to create a more caring, equal society, requires people to want to change the way in which the world works; and if they are to change something, they must first understand how it works at the present time. In my view, this is a second major applied educational task, which involves emancipation — making people aware of how society operates, so that they can take control of it.

This goal of emancipation requires education that goes beyond both the empirical level and the study of what people believe about the world, and reaches the heart of society itself. That project is now widely termed realism (see Johnston 1986a). It is built on the following argument:

1. In any mode of production, there are certain imperatives that form the driving force. If these are not allowed to operate, then a crisis (in the true sense of that word) is faced.
2. Such imperatives cannot be apprehended empirically since they are abstract forces rather than phenomena, though clearly phenomena reflect the forces. Nor can they be apprehended reflectively, in the way that humanistic research is conducted (Johnston 1986b), because access to people's thoughts and how they interpret the world does not tell us much about the nature of the world in which they work — only their view of it.
3. Such imperatives are deterministic at the general, or theoretical, level, but not at the empirical. Thus, although accumulation through the making of profits is part of the dynamo of capitalism, how profits are to be made is a product of individual decisions. People interpret both how capitalism works and how best to make it work in a particular context.
4. Since context is crucial to the empirical working-out of capitalism, it follows (from the discussion about learning places in the preceding section) that people in different places may decide to put capital to work in different ways and that others, those who sell their labour to capital, may respond in different ways too.

Two conclusions can be drawn from this argument. First, one needs to go beyond the empirical appearances to appreciate what capitalism really is, because of the immense variety of interpretations of capitalism that we have on the earth's surface. Second, geography is not necessary to capitalism but is a highly likely element of it. (Capitalism also has a geography because of the way in which space is used, as Harvey [1982] cogently argues. See also Smith 1984; 1986; and Browett 1984; 1987).

To understand what capitalism is, therefore, we have to go beyond the empirical representations of capitalism in action and the personal interpretations of action in capitalism. This involves the development of theory, of abstract conceptions which lay bare the forces that drive society. Such an activity could be entirely abstract, a set of axioms and theorems which are laid out in terms that have no apparent connection with empirical appearances, but this would be absurd for the following reasons.

First, capitalism is not a static mode of production; indeed change is necessary to its continued existence (see Lash and Urry 1987). Although the fundamental forces are unchanging, because they contain within themselves antithetical elements (the seeds of their own destruction), it is necessary for the people who operate capitalism to develop new strategies for coping with crisis. As our daily experience shows us, the inventiveness that people display produces a great variety of changing strategies and structures, not least changing geographies.

Second, the goal of a realist study of capitalism (or of any other mode of production) is to help people to appreciate the forces that produce its empirical complexity so that they can, if they wish, come to grips with those forces (or else evaluate the political programmes that are presented to them as being designed to counter, if not replace, those forces) and produce a different situation.

Thus the theoretical work must always keep in touch with the empirical world, because changes in that world pose new questions and the outcome of understanding those changes might be to propose others.

The type of education for emancipation that I am proposing in this section, therefore, involves people not only in coming to understand themselves and others but also in understanding the context in which they all live, so that they can take control of that context. Such education takes the empirical world as both its starting-point and its ending. Since that world is forever changing, so the educational process must at least keep pace. And because the world is spatially variable, since capitalism is not the same set of actions repeated many times in many places but rather a massive number of different actions conceived and executed in different places but with the same ultimate goals, studying the geography of capitalism is a necessary element in promoting its understanding.

Again, then, we see a utilitarian case for regional geography, for studying the spatial variety of places. It is not an end in itself (as Harvey [1987] makes clear in his critique of recent Marxian work by British geographers) but a means to an end. The goal of regional geography (or locality studies, which is the popular term) is not just to portray the peculiarities of places *x* and *y* but is to enlighten attempts to understand the dynamic of capitalism and thereby give people the power to decide the future of that dynamic.

Method

Having discussed why we need to study regions (or places, or localities, or locales ... I do not want to become embroiled in those semantic issues), I turn now to how we go about such study. In order to focus the argument, I want to start with a list of expectations, of what it is that regional study should demonstrate.

1. That the creation of regions is a social act. Regions differ because people have made them so. Undoubtedly in many cases differences in the physical environment will influence the creation of regional variations, with different environmental conditions stimulating different individual and social responses. (This is particularly so in very localized pre-capitalist modes of production, under which survival is very much tied to reaching a viable accommodation with the physical environment.) But similar physical environments can be associated with very different human responses, and similar patterns of spatial organization can be found in very different milieux.

2. That regions are self-reproducing entities, because of their importance as contexts for learning and socialization. As milieux, they provide the role models for self-development, as well as the contexts for the nurturing of particular sets of beliefs and attitudes. (Residential segregation in cities is a good example of the former; the propagation of a national ideology illustrates the latter.) Thus people are made in places: if places differ, then people will too.

3. That the self-reproducing characteristics of regions are not deterministic, so that culture does not have an existence as a thing-in-itself, separate from the individuals who make and remake it (see Duncan 1980). Aspects of the characteristics of a region can change as a consequence of:

(a) The need to respond to new stimuli — whether in the physical or the human environment (a tornado, perhaps, or the introduction of a new form of transport);

(b) Contact with people from other regions, and the adoption (in part or in whole; voluntarily or forced) of some of their ways;

(c) The determined actions of individuals within the region to promote changes in order to advance their own interests (i.e. through the exercise of power).

Thus, just as people learn individually in a place which reproduces its characteristics as they do, so people learn collectively, altering their culture as they do. That they must learn is necessary; how they should respond as they learn is not, so that the future nature of a region (indeed, even if it is to remain a separately identifiable region) cannot be predicted, since we cannot be sure what aspects of the present (and hence of the past) will be retained for the future, what new features will be introduced, or what hybrids will be formed. Change is a human creation. (People make their own history, but not in conditions of their own choosing!)

4. That regions in a capitalist world-economy are not isolated, independent units whose residents can control their own destinies. The world has always been an interdependent physical system: it is now an interdependent human system too. (Clearly there are still a few precapitalist remnants that provide exceptions to that rule, but they are of trivial importance to the system as a whole. The links between the capitalist and socialist states are more problematical, but are not of particular relevance here.) This interdependence has become very much greater in recent decades, as we are well aware, with the consequence — as made very clear in Peter Taylor's (1985) seminal text on political geography — that people in places are subject to the dynamics of an essentially placeless economic system, the capitalist global world-economy. They may apprehend that system in its local manifestations alone (at what Taylor [1982] calls the *scale of experience*), but they will understand the reasons for those local manifestations only through appreciation of the entire system (what he calls the *scale of reality*). Emancipation, then, must come from setting the region in context.

5. That regions are not simply the outcome of human activities, so that the geographical mosaic is studied merely

as a consequence of local responses to global imperatives. Regions are deliberately created, by individuals and institutions that exercise power in society, to promote certain goals. Those regions may be indeterminate in extent, in that the creators pay no particular attention to boundaries, but they may be formally defined, with demarcated and defended boundaries.

By far the clearest example of the creation of formal regions in that sense is the state, a region with its own characteristics, which is necessarily territorial (see Mann 1984) and which is used by powerful interests to promote certain goals. One of the characteristics of most states is the attempted creation of a national identity, a sense of belonging: residents of the state are defined as 'us', and those outside as 'them', with all the stereotyping that is consistent with the withdrawal and coping strategies described earlier. (Such stereotyping is frequently negative but not necessarily so; the residents of some other states may be presented as 'our friends', at least temporarily.) Thus, to use Taylor's terminology again, the state is the *scale of ideology*, designed as a coping strategy within the complex, interdependent global system.

Although the national state provides an excellent example of a region which is created in order to promote particular interests, it is by no means the only one. In some situations — as I have described in work on American suburbs (Johnston 1984b) — local government units can be similarly employed, and there are many examples from around the world of the creation of closed residential communities for such purposes. They all illustrate the importance of space as a resource to be manipulated, in the creation, recreation, and restructuring of regions within which people are made.

6. That regions are not only containers in which to live separate existences but also the potential sources of conflict. Capitalism is an economic system which is built on conflict, as well as on competition, but the existence of regions, of separately defined territorial units within that system, enhances the potential for conflict. This is clearly most dangerous at the state scale, because states have a virtual monopoly on legalized violence, and because the interpretations of those who wield state power, that the residents of the other states are 'threats', have led to the apparently unstoppable escalation in the creation of weapons of mass destruction. They have created the situation that is so geographically defined by Bunge (1985):

'Our planet is big enough for peace but too small for war'.

Although the potential for conflict and its consequences is greatest at the state level, we must realize that it is present at other scales too. This is the thesis developed by Sennett (1973; see also Johnston 1985; 1989), who argues that the withdrawal and stereotyping strategies sustain what he describes as an 'adolescent immaturity'. Separation and mutual misunderstanding/mistrust provide a basis for conflict and prevent a coming together that will produce mature accommodation to cultural differences. Thus regional geography, at all scales, provides the necessary conditions for interpersonal and inter-group conflict, which may or may not end in violence (Johnston, O'Loughlin, and Taylor 1987).

Having set out these six expectations for regional geography, the final question that I must pose is: how can they be realized? I wish to argue that realization will not come from a separate subdiscipline of regional geography; rather, I contend that what we need is a regional perspective that informs all geographical activity. (In this way, I extend my argument against certain fixations in the study of geography: Johnston 1986a; 1986c.)

This contention may seem to be far-fetched and calling for a break with traditional practice that has little chance of success: the expectations are too great and the methodology ill-defined. The goal that I set, through that list of expectations, is to understand the world as a whole — i.e. emancipation — by using its parts as the empirical means. Indeed so, but the integration that is supposed to occur rarely, if ever, materializes, even in curricula that seek to move from whole to part to whole again. I prefer to avoid the possibility that the parts will be seen as more important than the whole, and that the crucial general goal — holistic appreciation and emancipation — will be lost. I admit that my argument is tendentious and tentative but, given our failure to achieve emancipation to date, I am prepared to gamble on something new. (I accept, of course, the need for an empirical base — what is where — but assume that it is already in place.) What I propose, therefore, is the development of a perspective within all geographical study, rather than promotion of one of its current elements.

Region in geography rather than regional geography

One of the clear realizations of several geographers in recent years is that the disaggregation of geography into its component parts (urban and rural, economic and social, agricultural and industrial, etc.) is counter-productive, because it tears apart wholes that must be retained intact. (At a different scale, the same can be said of the division of the social sciences: society should be considered as a whole. This has led to calls for the de-definition of geography: Eliot Hurst 1980; 1985. For the reasons outlined here, I do not subscribe fully to that case. A geographical perspective is a valid, even vital, part of social science education, given the present construction of educational institutions and academic disciplines, but only if that perspective is set in a holistic context.) This does not necessarily mean that those component parts of geography should be abolished, since they provide necessary foci of attention on certain phenomena. But those foci must not be myopic and/or blinkered; the part that is the centre of attention must be studied as an element of an interacting whole. (The part could be derived from systematic studies — industry, perhaps, or shopping centres — or it might be a particular area. If it is the former, the systematic part must be set in holistic context with regard to other parts; if it is the latter, the particular area — and perhaps some systematic element of it — must be seen as part of a wider world-system.)

What is that interacting whole? It is tempting to call it the regional culture except that, as Williams (1976) points out, culture is already used in a whole range of ways, though it is usually employed to refer to certain activities that comprise 'the works and practices of intellectual and especially artistic activity' (p.80: see also Williams 1981). Further, there is already a well-established subdiscipline of cultural geography (very much a creation of the United States within the English-speaking world) which deviates somewhat from Williams's focus with its concentration on artifacts as landscape elements, plus major forms of discourse, such as languages, and modes of social organization, such as religions. (Some of its work appears to focus on the relatively trivial and esoteric, rather than on the central elements of social organization: see Rooney, Zelinsky, and Louder 1982). Perhaps I should try to revive Whittlesey's (1956) term compage, a unified complex of diverse elements. Or perhaps the terminology is irrelevant.

What is important is to realize that much behaviour is local not only in its impact but also in its provenance and that such behaviour — whether social, economic, or political in terms of the three main systematic subdisciplines of human geography — is strongly influenced by its local context. Therefore what we need is a framework for looking at that context. In the broadest sense, places are made up of three components: a physical environment, a built environment (i.e. a pattern of land uses), and people. Here I want to focus only on the third of those interrelated components, the people, and suggest (developing on the ideas of Leeds 1984) that such a framework is four-dimensional, that we should define a region (place, locale ...) on four criteria.

Place in the spatial division of labour

In the global economic system, each region occupies a particular niche; it plays a particular role in the system's operation. It may (depending especially on spatial scale) be a unitary role — a dairy farm, perhaps, or a car factory — but it is more likely to be made up of a combination of roles, as often described in factorial ecologies of census and other data.

The social relations of production

Whatever its role(s) in the global economic system, the region will have a particular (or typical) way in which that role is organized. The most important element of that characterization is the way in which work is organized. At the coarsest scale, we can distinguish between, for example, slave, feudal, communistic, and capitalist societies. In more detail, we can look at how work is organized within each because, as we are frequently told today with regard to contrasts between British and Japanese working practices (and as the earlier discussion has indicated), there are many ways in which capitalism can be put into operation as a mode of production that is built on the sale and purchase of labour power.

The institutional structure of formal organizations

A materialist view of society sees production, or work, as its fundamental component, and it is how production is organized that characterizes the society (hence the Marxist term mode of production). But our understanding of capitalism shows us that the economic structure is insufficient to sustain it unless it is buttressed by others, notably a political structure: the state is necessary to capitalism (see Mann 1984). Thus each place has not only its social relations of production but also its political relations: and again, the theoretical necessity does not require

a particular empirical form. Hence there are political variations between regions, differences in interpretations of not so
much what the state should do as how it should do it and what
form it should take.

The informal structure of society

Just as an economic system needs a supporting political
structure so it also requires a social structure, a way in which
reproduction is organized and which provides a context for
life outside the workplace (which could be called culture).
That social structure includes the way in which daily life is
organized (in family units in our case at the present time, but
with many varieties of family structure: Todd 1985; 1987) but
encompasses many other aspects of society: gender relationships; religions; ethnic relationships; and so on.

These four aspects of society are clearly interrelated
empirically. (The political structure may be based in part on
the religious structure; the social relations of production may
reflect the ethnic structure.) By identifying them separately,
however, I am providing both a way of disaggregating a
regional society and a perspective on its major components.
The stress, however must be on the whole, on the links among
those components.
     There is no room here for the presentation of detailed case
studies to support any argument (which has been done
elsewhere: see Johnston 1986d; 1986e). All I can stress is that
one cannot understand a part of a region without understanding its whole. The focus will often be on one component of the
region only, one of the dimensions of my four-fold framework, but any attempt at understanding the one part from the
other three will be at best partial and at worst misleading.
Certainly, any attempt at emancipation, at uncovering the real
forces driving society, will achieve little if it abstracts parts
from their whole.
     It is for that reason more than any other that my case for
regional geography is not an argument for reconstituting a part
of the discipline that focuses — region by region — on the
features of defined segments of the earth's surface. My case
is that all human geography has to be regional geography, in
that all attempts to appreciate particular phenomena must
realize that they are parts of complex wholes, interacting
within those wholes to reproduce and change them.

In summary

This chapter has deliberately been written in a contentious way, to provoke reflection and debate. A regional perspective is necessary to geographical (indeed to social scientific) inquiry, not just to improve geography but rather and far more importantly — to tackle the task of ensuring that the future has a geography. The term crisis is undoubtedly overused at the present time, and it is little more than facile to describe many situations that we face as crises (Johnston and Taylor 1985). But a potential crisis is staring us in the face, a crisis which we can easily bring about by a short chain of decisions that could lead to massive, if not complete, destruction of ourselves and our means of survival. If we can contribute to ensuring that such a crisis never eventuates, then we will have proved the value of our discipline.

That value will be demonstrated by the educational rather than the technical content of geography, by its development as a means, first, of promoting self- and mutual understanding and thereby reducing conflict potential, and, second, of emancipating people, of bringing them to an understanding of how their world works. Regions, as I have argued here, are crucial components of the working of the world. Our task is to promote that argument, by promoting the regional perspective. In that way, we will enable people to connect the various parts of their local experience and to connect their region with not only other regions but also the forces that underlie the creation of all regions. For that, we do not need regional geography, but we do need regions in geography.

References

Browett, J.G. (1984) 'On the necessity and inevitability of uneven spatial development', *International Journal of Urban and Regional Research* 8:155-76.

— (1987) 'On the significance of geographical space: reply to Smith', *International Journal of Urban and Regional Research* 11:262-79.

Bunge, W. (1985) 'Epilogue: our planet is big enough for peace but too small for war', pp.289-91 in R.J. Johnston and P.J. Taylor (eds) *A World in Crisis?: Geographical Perspectives*, Oxford: Basil Blackwell.

Daniels, S. (1985) 'Arguments for a humanistic geography', pp.143-58 in R.J. Johnston (ed.) *The Future of Geography*, London: Methuen.

Duncan, J.S. (1980) 'The superorganic in American cultural geography', *Annals of the Association of American Geographers* 70:181-98.

Eliot Hurst, M.E. (1980) 'Geography, social science and society: towards a

de-definition', *Australian Geographical Studies* 18:3-21.
— (1985) 'Geography has neither existence nor future', pp.59-91 in R.J. Johnston (ed.) *The Future of Geography*, London: Methuen.
Freeman, T.W. (1961) *One Hundred Years of Geography*, London: Duckworth.
Gould, P.R. (1979) 'Geography 1957-1977: the Augean period', *Annals of the Association of American Geographers* 69:139-51.
Guelke, L. (1974) 'An idealist alternative in human geography', *Annals of the Association of American Geographers* 64:193-202.
Hart, J.F (1982) 'The highest form of the geographer's art', *Annals of the Association of American Geographers* 72:1-29.
Harvey, D. (1982) *The Limits to Capital*, Oxford: Basil Blackwell.
— (1987) 'Three myths in search of a reality in urban studies', *Environment and Planning D: Society and Space* 6.
Johnston, R.J. (1984a) 'The region in twentieth century British geography', *History of Geography Newsletter* 4, pp.26-35.
— (1984b) *Residential Segregation, the State and Constitutional Conflict in American Urban Areas*, Academic Press, London.
— (1985) 'To the ends of the earth', pp.326-38 in R.J. Johnston (ed.) *The Future of Geography*, London: Methuen.
— (1986a) *On Human Geography*, Oxford: Basil Blackwell.
— (1986b) *Philosophy and Human Geography* 2nd ed., London: Edward Arnold.
— (1986c) 'Four fixations and the quest for unity in geography', *Transactions, Institute of British Geographers* NS11:449-53.
— (1986d) 'Placing politics', *Political Geography Quarterly* 5:S63-S78.
— (1986e) 'Job markets and housing markets in the "developed world"', *Tijdschrift voor Economische en Sociale Geografie* 77:328-35.
— (1989) 'People and places in the behavioural environment', pp. 235-52 in F.W. Boal and D.N. Livingstone (eds) *The Behavioural Environment: Essays in Reflection, Application and Criticism*, London: Routledge.
— O'Loughlin, J., and Taylor, P.J. (1987) 'The geography of violence and premature death: a world-systems view', pp.241-59 in R. Vayrynen (ed.) *The Quest for Peace*, London: Sage.
— and Taylor, P.J. (1985) 'Introduction: a world in crisis?', pp.1-11 in R.J. Johnston and P.J. Taylor (eds) *A World in Crisis?: Geographical Perspectives*, Oxford: Basil Blackwell.
Lash, S. and Urry, J. (1987) *The End of Organized Capitalism*, Polity Press, Oxford.
Leeds, A. (1984) 'Cities and countryside in anthropology', pp.291-312 in L. Rodwin and R.M. Hollister (eds) *Cities of the Mind*, New York: Plenum Press.
Mann, M. (1984) 'The autonomous power of the state: its origins, mechanisms and results', *European Journal of Sociology* 25:185-213.
Rooney, J.F., Zelinsky, W., and Louder, D.R. (1982) *This Remarkable Continent*, College Station, Texas: Texas A and M Press.

Sack, R.D. (1983) 'Human territoriality: a theory', *Annals of the Association of American Geographers*, 73:55-74.

— (1986) *Human Territoriality: Its Theory and History*, Cambridge University Press, Cambridge.

Sennett, R. (1973) *The Uses of Disorder*, Harmondsworth: Penguin Books.

Shevky, E. and Bell, W. (1955) *Social Area Analysis*, Stanford: Stanford University Press.

Smith, N. (1984) *Uneven Development*, Oxford: Basil Blackwell.

— (1986) 'On the necessity of uneven development', *International Journal of Urban and Regional Research* 11:87-103.

Taylor, P.J. (1982) 'A materialist framework for political geography', *Transactions, Institute of British Geographers* NS7:15-34.

— (1985) *Political Geography: World Economy, Nation-State and Locality*, London: Longman.

Todd, E. (1985) *The Explanation of Ideology* Oxford: Basil Blackwell.

— (1987) *The Causes of Progress*, Oxford: Basil Blackwell.

Whittlesey, D. (1956) 'Southern Rhodesia: an African compage', *Annals of the Association of American Geographers*, 46:1-97.

Williams, R. (1976) *Keywords*, London: Fontana.

— (1983) *Culture*, London: Fontana.

9 Metamorphosis: how spatial facts change into classes of geographical regions

Gerda Hoekveld-Meijer

Introduction

Consider the following items: Swiss motorable passes, Swiss cantons, and Switzerland. How will they be treated by a spatial analyst, a thematic geographer, and a regional geographer?

The spatial analyst will study motorable passes in Switzerland, or in those cantons which have motorable passes, in relation to other localized phenomena in and outside Switzerland, or in and outside the cantons in question.

The thematic geographer will analyse the distribution of motorable passes as an attribute of Switzerland, or as an attribute of the cantons in which passes are located, or as an attribute of a pass region, the boundaries of which are drawn by the geographer. The thematic geographer will relate these motorable passes in their appearance of *attributes* with other attributes which are relevant to the former, in order to compare Switzerland, or some Swiss cantons, or Swiss pass regions with other political, administrative, or pass regions respectively.

The regional geographer will describe and analyse the pattern of areal differentiation as an attribute of Switzerland. This pattern consists of all Swiss cantons, each with a distribution of motorable passes as an *attribute*. In comparison with the thematic geographer, the regional geographer studies motorable passes on another level of analysis. The latter considers motorable passes as a class attribute of *all* Swiss cantons. Cantons without passes are included in the analysis. On the basis of the pattern that is formed by the variation of motorable passes per canton, Switzerland can be compared with other countries and with other differentiation patterns.

Having the same data at their disposal, the three scientists observe different things because they have different aims. For the spatial analyst, motorable passes are the object of study, whereas Switzerland or Swiss cantons are merely attributes of this spatial class. The thematic geographer perceives Switzerland or some Swiss cantons in their undivided form. As such they are the object of study. Number and distribution pattern of motorable passes are their attributes. For the regional geographer, the object of study is Switzerland as a region consisting of regions (i.e. all Swiss cantons). The cantons are its attribute and motorable passes the thematic attribute of the cantons. Obviously, regional geographers have to combine spatial and thematic geographical knowledge in order to create regional geographical knowledge. This knowledge is the result of several transformations in which members of a class of localized phenomena seem to lose their old attributes. By gaining new ones, the original members are reborn with other attributes.

In order to demonstrate these transformations and to reinforce the status of regional geography among the other subdisciplines of geography, we shall describe and relate those concepts which are instrumental in transforming spatial knowledge into regional geographical knowledge.

Spatial classes: the domain of spatial knowledge and spatial concepts

In order to understand a metamorphosis we have to know how the main concepts which rule the domain of spatial knowledge are mutually interrelated. They are the well-known terms: location, distance (and orientation), situation (and connection), spatial interaction, spatial distribution, and network. These concepts do not belong to the same level of analysis. Distribution and network are attributes of a spatial class that is studied as an object. The others are attributes of the members of spatial classes. Each member of a spatial class has two characteristics in common: each belongs to the same category or class of phenomena and each is located in the same 'space'. This space, however, is a 'real' place or region, known by its toponym. The region in which a spatial class is present can be the whole world, or part of it. For instance, there are motorable passes all over the world; there are motorable passes all over Switzerland. The frequency with which members of a class are present is a third attribute of the spatial class. In the case of Swiss motorable passes, it is 32 (and in the near future 33). Category and region determine the frequency of a class to be studied. We have selected, of all passes in the world, only motorable passes (a categorial selection) and, of all the motorable passes, only those which are located in Switzerland

Figure 9.1a Distribution of Swiss motorable passes

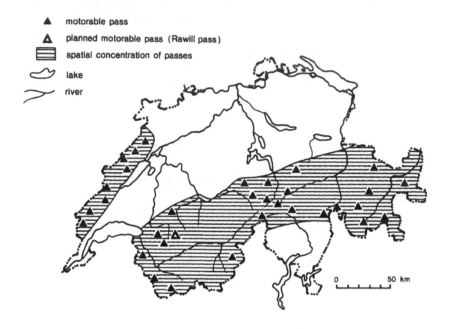

Figure 9.1b Networks of Swiss motorable passes

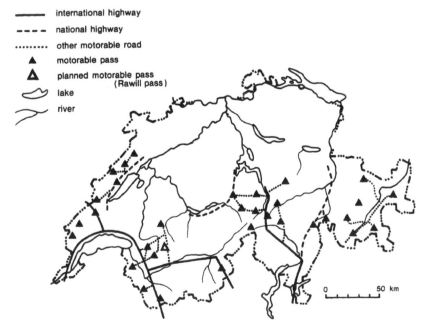

Source: Atlas der Schweiz, maps 70, 72

Figure 9.1c Areas of distribution of Swiss passes with regard to their situations

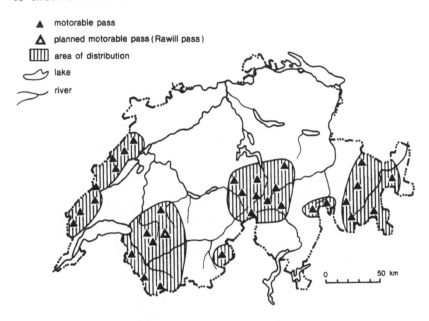

Figure 9.1d Interaction of lorries and foreign passenger cars between motorable passes and Swiss cantonal capitals

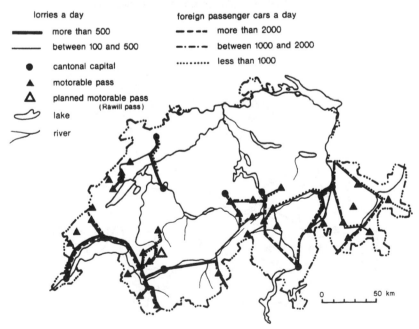

Source: Atlas der Schweiz, maps 70, 72

Figure 9.1e Areas of distribution of Swiss motorable passes based on the intensity of traffic of lorries and foreign passenger cars

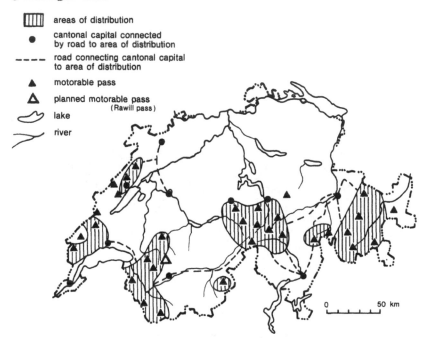

Figure 9.2a Distribution of Swiss cantonal capitals

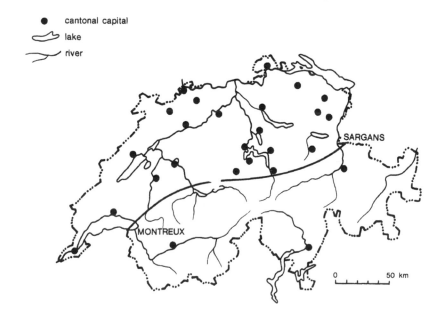

Figure 9.2b Networks of Swiss cantonal capitals

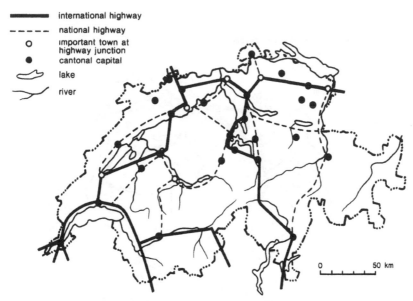

Source: Atlas der Schweiz, maps 70, 72

Figure 9.2c Areas of distribution of Swiss cantonal capitals situated at international and national highways

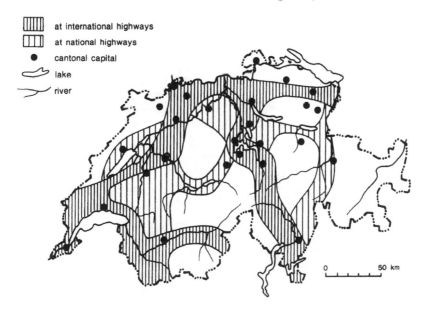

Figure 9.2d Interaction of lorries between Swiss cantonal capitals

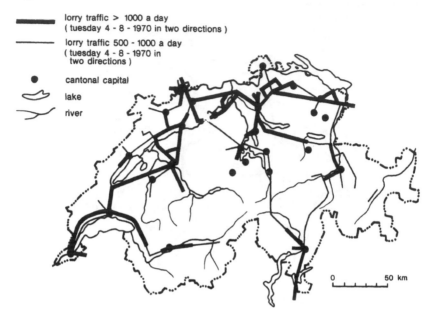

Figure 9.2e Areas of distribution of Swiss cantonal capitals based on intensity of lorry traffic

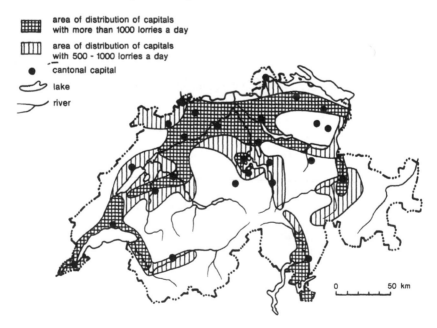

(a regional selection). Each pass has its unique location, its distance, and its orientation to other members of the same class, as well as its situation and connections and interactions with all other members of the same class. Apart from the relations between members of one and the same spatial class, each member can have a relation (distance, situation, and interaction) with members of another class in the same region (for instance, Swiss motorable passes and Swiss cantonal capitals); with members of the same category which are located in another region (for instance, Swiss passes and Austrian passes); and with members of another category in another region (for instance, Swiss motorable passes and Italian regional capitals). Because of internal relationships, which include distances (always in combination with orientation), connections, and interactions between members of the same spatial class, the class that is studied as the object has the following attributes: a distribution pattern, a network, and an interaction pattern respectively. Relations existing between members of one spatial class and members of other classes in and out of the space of that particular class do not belong to the patterns of this class.

In Figure 9.1a, the distribution pattern of Swiss motorable passes appears as two concentrations: one between the lines Montreux-Sargans and Simplon-San Bernardino and the other between the Swiss-French border area and the western Swiss lakes. In Figure 9.2a, the distribution pattern of Swiss cantonal capitals is best described as one concentration in the northern half of the country, north of the line Montreux-Sargans. In Figures 9.1b and 9.2b, there are several networks. The first map shows that some passes are indirectly connected to each other by highways, usually open throughout the year. There are other Swiss passes which are connected only through main roads which are often closed between September and April/May. In Figure 9.2b, we notice that most of the Swiss cantonal capitals are connected through international and/or national highways. The areas of distribution of the subclasses of Swiss motorable passes and cantonal capitals are drawn in Figures 9.1c and 9.2c respectively. Both classifications and corresponding areas of distribution are based on the variable class member attribute 'situation'. Each separate area of distribution reflects the shape of the network of a subclass. As such, 'network' is the variable attribute of the class. Comparing these maps, we see that the areas of distribution of motorable passes remain within the broad zones that are indicated in Figure 9.1a. However, the areas of distribution of Swiss cantonal capitals entirely break up the neat north-south partition of Switzerland. The interaction pattern of Swiss motorable passes, based on streams of lorries and foreign passenger cars, is shown in Figure 9.1d. The nearest cantonal capitals are integrated in this pattern. Figure 9.2d displays the flow of goods that is transported by either 500-1000 lorries or more than 1000 lorries per day passing cantonal capitals. Taking into consideration the traffic patterns towards and between motorable passes and cantonal capitals, we have delineated areas of distribution of motorable passes (Figure 9.1e) and cantonal capitals (Figure 9.2e). Figure 9.1e indicates five clusters of Swiss passes, one situated in the core, one in Graubünden, one in western Wallis, and two in the Swiss-French border area. The areas of distribution of Swiss cantonal capitals, which are connected through intensive lorry traffic (see Figure 9.2e), appear as an undulating zone running from Lausanne via Basel to St Gallen and further east to St Margarethen on the Swiss-Austrian border and with an extension to the west via Delémont. The capitals in the upper valleys of the Rhône, Ticino, and Rhine all have their own narrow area of distribution because they all have a relatively large flow of traffic passing through.

In an ideal situation, networks and interaction should have the same pattern, implying that they really coincide, and that, consequently, their corresponding areas of distribution overlap. We have seen that such an ideal situation exists in the case of the Swiss motorable passes (cf. Figures 9.1c and 9.1e). However, networks and interaction between cantonal capitals only partially coincide. Comparing Figures 9.2c and 9.2e, we might conclude that some areas of distribution had an over-developed network in relation to interaction patterns which were measured in 1970 and, especially, that the Central Swiss pass area and the area between Thun-Montreux-Lausanne had oversized networks. Other areas of distribution lacked adequate networks, for instance the area around Delémont and to some extent a small zone immediately east of Chur.

All of the methodological concepts that have just been mentioned are interrelated, as is indicated in Figure 9.3. Here we see the first metamorphosis: a class with invisible and abstract attributes emerges as an object with concrete attributes. This transformation is possible thanks to the locational attributes of all of the members of the spatial class. No matter how many locations and distances there are, we nevertheless perceive only one distribution pattern. However, this is entirely determined by the frequency and location of the members of the spatial class and their mutual distances. After we collect data on quality and quantity of flows (exchange relations or interaction) between all located members of the class, we are able to draw an internal inter-action pattern which depends on perceived differences in quality between the members of the same class. Quality is a categorical attribute with the connotation of value judgement. Difference in quality is a relational, categorical concept, with the same connotation. Differences in qualities, resulting in or being the result of interaction, can best be described by the term complementarity, a concept which is used as a locational attribute. Perception of complementarity varies in different historical periods and in different cultures.

Through this concept, history and culture, and their corresponding time-space relations, enter the otherwise strictly methodological conceptual structure of Figure 9.3.

When we trace all existing connections between located members of the class, we discover an internal network which includes either all or only some members. The form of the network depends, among other things, on the internal inter-action patterns which existed in the past and on more recent developments in exchange patterns. On the basis of the form of the network and/or interaction patterns, we are allowed to delineate the area(s) of a distribution. They are the second type of space in relation to the first selected space (region 1) of the spatial class. In case the distribution is evenly spread over this region and the network includes all members, the area of distribution equals region 1, or the 'space' of the class. In case there are two or more networks to be distinguished, there are two or more areas of distribution in which the members of the class are clustered through their locations and connections. These area(s) of distribution either cover or do not cover the entire region. In case part of the region does not belong to one area of distribution, as is the case in Figures 9.1e and 9.2e, this area is of no importance to the spatial class. However, this particular area is of the utmost importance in case the spatial class is studied not as an object but as a property of region 1, as we shall see later on. For the time being we merely draw attention to the fact that, by jumping between the levels of analysis of a spatial class, we stand a good chance of ending up with more spatial classes than we began with. After starting with Swiss passes, we detected Central Swiss passes, Graubündner passes, Walliser passes, and Swiss-French border-region passes (see Figures 9.1c and 9.1e).

When we decide to consider a spatial class as an object, it can be analysed as a member of another spatial class. There-fore the question is: to which 'space' and to which 'spatial' class does the object 'Swiss motorable passes' belong?

Figure 9.3 Interrelated concepts on three levels of analysis applicable to spatial classes

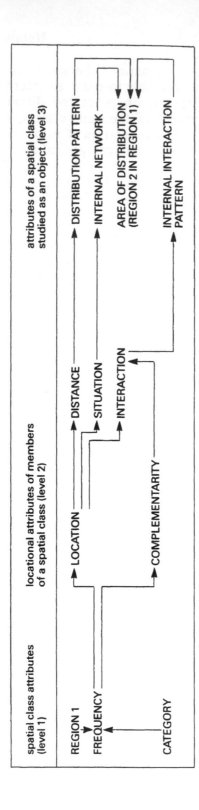

This new class consists of spatial classes either in the same 'space' (region 1) or in another 'space' outside the boundaries of this region.

Spatial classes sharing the same space

If we study spatial classes within the same region 1, that is, in the same 'space' (for instance, Swiss non-motorable passes, Swiss cantonal industrial towns, Swiss tourist centres, etc.), these classes differ according to category. Since comparisons between categories are not possible with regard to their ordinary, categorical attributes, only their spatial and locational attributes can be studied. Therefore we can compare their distribution patterns, networks, and interaction patterns as attributes of a spatial class which consists of these special classes sharing the same 'space'. The selection of the spatial classes to be studied as members of such a spatial class depends upon the hypothetical or theoretical relations between the elements of the distribution that is to be studied. Whether the assumptions are sound or not, the results of such an analysis are first, a description of the areal overlap of the various areas of distribution; second, a description of how the networks and interaction patterns of the distributions, which consist of objects belonging to different categories, are joined; third a determination and delineation of areal associations which consist of spatial phenomena that belong to different categories and are connected to each other through networks and interaction patterns. Each areal association is theoretically founded, because the idea of comparing the distributions of categorically different classes, such as Swiss passes and cantonal capitals, originated in a hypothetical relationship between the members of these classes. This may be expressed in the following hypothesis: the looser the connections between a capital and the points where important international and national roads cross natural and/or political borders, the weaker its economic position and the smaller its cross-cultural base, as might be indicated by exchange flows of goods and ideas between capitals. In case networks or interaction patterns of Swiss cantonal capitals and motorable passes are not joined, the assumptions are proved to be wrong for these particular spatial classes in their particular space and the areal overlap, which might be visible on a map, is not a real areal association. It is just a coincidence without direct meaning for the class members.

Areal association has various appearances depending on

Figure 9.4a Areal associations of Swiss cantonal capitals and motorable passes based on their international networks

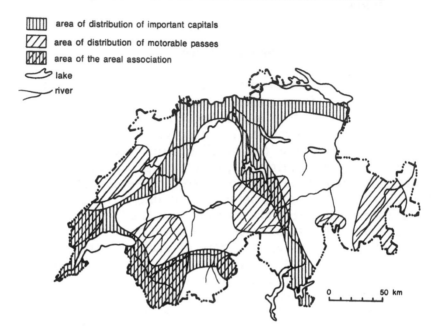

Figure 9.4b Areal associations of Swiss cantonal capitals and motorable passes based on their interaction patterns

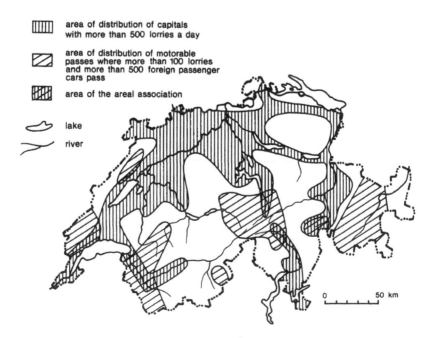

way areal distributions overlap. First, it is possible that the area of the association consists of members of the two or more spatial classes involved, each with its own network and interaction pattern. This variant of areal association is usually called a spatial system. Second, it is also possible that not one of the constituent class members of the association occurs in the area of the areal association, but only parts of their networks and/or interaction patterns. Third, yet another possible variant is the situation in which the area of the association contains at least one (member) of the associated classes and parts of their networks and/or interaction patterns.

The areal associations in Figure 9.4a are based on Figures 9.1c and 9.2c. They represent the third type. The Rhône valley association in Figure 9.4b, which is based on Figures 9.1e and 9.2e, belongs to the second category. The Rhine (anterior) valley association, again in Figure 9.4b, includes all of the passes of the association (San Bernardino and the Splügen) but excludes the capital Chur. As such it also belongs to the third variant.

Ideally, the areas of the associations that are based on networks and interaction patterns between members of different spatial classes cover the same area. Comparing Figures 9.4a and 9.4b, we see that the Swiss situation is far from ideal, because only the main valleys in the areas of the interlinked passes also participate in the cantonal capital interaction pattern. When members of one or different categories are linked through networks and any form of interaction, we usually call this phenomenon a spatial system. There are two variants of spatial systems: the homogeneous and the heterogeneous. The heterogeneous system is an attribute of a class which consists of spatial classes sharing the same space. The characteristic of this 'class of classes' is expressed in the way in which all of the members of the categorically different classes are related or not related. In Switzerland we created such a spatial superclass just by declaring two other spatial classes as its members: Swiss motorable passes and Swiss cantonal capitals. The variable attributes of this superclass are the number and character of the spatial subsystems that are clusters of categorically different objects which are connected more to each other than to the other objects of the two spatial classes. Figure 9.5 shows how areal association and spatial system, as attributes of a spatial (super)class consisting of spatial classes within the same space, depend on the interlinkages between networks and interaction patterns. It also indicates how these attributes of the members of categorically different spatial classes are

Figure 9.5 Interrelated concepts on two levels of analysis which are applicable to spatial classes consisting of spatial classes within the same region

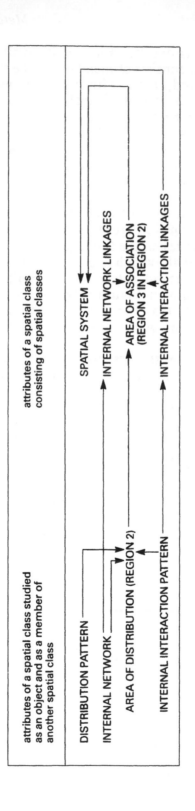

related to the attributes of a spatial (super)class which consists of these classes but is studied as *one* object.

Spatial classes in different spaces

When we study the spatial class, Swiss passes, as a member of the spatial class, European Alpine passes, the attributes of all of the Swiss passes as one object can be compared with the attributes of the other European Alpine passes. As soon as spatial classes which belong to different spaces are studied as objects, they all belong to one particular space or region, which includes all 'spaces' of all spatially different classes. This space is the fourth space to be analysed in relation to any first-selected space or region 1 (the second and third being areas of distribution and areas of areal associations respectively). It is not correct to compare only Swiss and Austrian passes. We should first establish to which spatial superclass (or region 4) the Swiss and Austrian passes belong in order to find all of the members with which the attributes should be compared. For instance, all spatial classes of passes in the European Alpine region are Swiss, Austrian, French, Italian, Yugoslavian, Romanian, Czechoslovakian, Bulgarian, Greek, and Spanish passes. Apart from their spatial attributes, distribution pattern, network, interaction pattern, and area(s) of distribution, a spatial class, which is observed as a member of a spatial superclass, automatically has all of the locational attributes belonging to any member of a spatial class (cf. Figure 9.3). This means that each spatial class is reduced to one or a few points, depending on its number of areas of distribution, although we have to keep in mind (literally) that these areas are filled with networks, interactions, and objects. This is shown in Figure 9.6 which is an extension of Figure 9.3. We can see that the concept complementarity is expressed in terms of the interrelated concepts of area of distribution, internal networks, and internal interaction patterns. These interrelated attributes are the basis for comparison and analysis of spatial classes which are observed as members of a spatial superclass.

People tend to experience these complementary 'filled spaces' of region 1 as places. This experience is most evidently expressed when they identify filled space by giving it a name or toponym. For a longer or shorter period these toponyms cover the exact areas of distribution or association. The Graubünden pass area was apparently experienced as a place, because it was given a name and is known as Engadin. The

Figure 9.6 Interrelated concepts on three levels of analysis which are applicable to a spatial class consisting of spatial classes which are spatially different

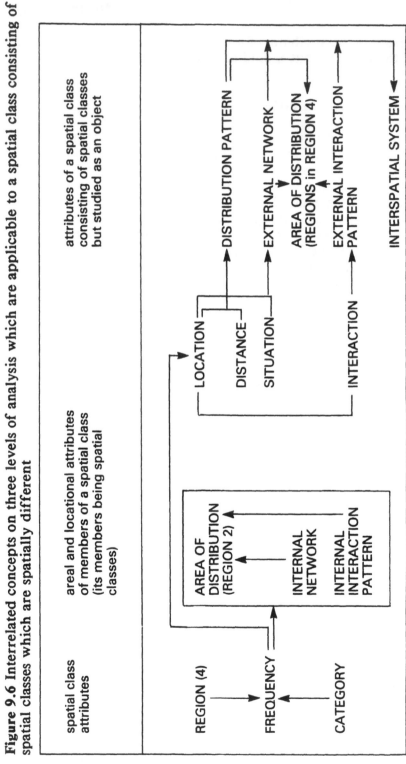

Figure 9.7a Subsequent metamorphoses of spatial facts within the first selected space (region 1) (to start a new run at least one new spatial class must be selected)

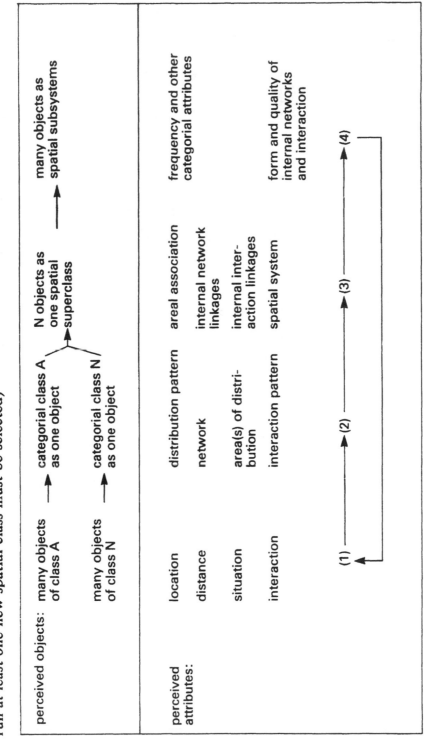

Figure 9.7b Subsequent metamorphoses within the fourth space (region 4) (external in relation to region 1 and internal in relation to region 4; to start a new run one new space including region 4 must be selected)

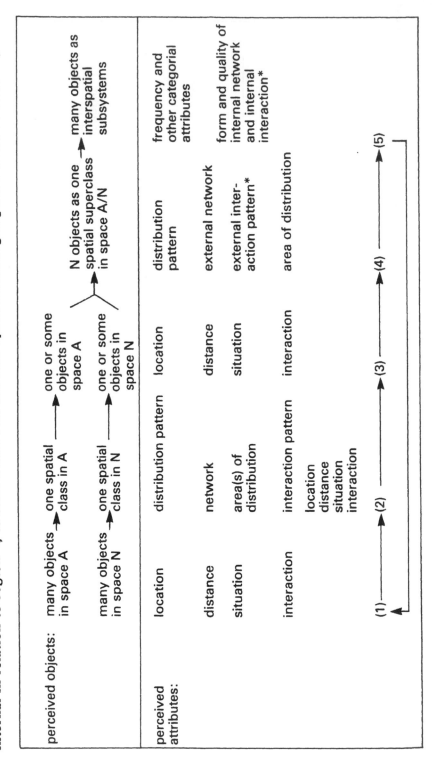

Central Swiss pass system not only was given the name Uri but also became the first Swiss canton. Thus, when we are studying spatial classes in a first and fourth space, we are witnessing a metamorphosis in region 1, immediately followed by another in region 4. Some or all members of one spatial class are transformed in a subregion of region 1, whereas this subregion is treated as a point in region 4, which includes region 1. All of these 'points' together have their own area of distribution, network, and interaction patterns which are, however, external in relation to their first space. They are the attributes of the spatial superclass of which all of these spatial classes (Swiss, Austrian, French, etc. passes) are members. These members might belong to the same category (passes) or to different categories (regional capitals, tourist centres). Whatever might be the case, they can be analysed as members of one class because they share one space. We can compare and analyse their distribution patterns, internal networks, and internal interaction patterns as exponents of their locations as well as of their distances, situations, and interaction patterns. In doing so, the spatial class of the 'fourth' space receives its attributes in the form of a distribution pattern of spatial superclasses, internal network(s), and interaction patterns between these classes and area(s) of distribution — which are, however, external in relation to the first-selected spaces — and 'inter' spatial subsystems, which are formed by the distribution patterns, networks, and interaction patterns. In a study in which the European Alpine region is the selected 'fourth' space of Swiss passes, the area(s) of distribution of all European Alpine passes is not the European Alpine region. Rather, it is that area in which the areas of distribution of most of the Swiss, Austrian, French, etc. passes, which are observed as members of spatial classes which are themselves spatial classes, are part of a network and interaction pattern (see Figure 9.6). These areas of distribution are the fifth type of area to be reckoned with in any study in which Swiss passes are observed as members and as an object. Some of these areas of distribution are more connected than other areas. As such, they form clusters which we can describe and analyse as individual interspatial subsystems within one space (i.e. region 4).

We have now reached the stage in our discussion where all of the ingredients are available to demonstrate two sets of geographical metamorphoses. They are ordered in Figures 9.7a and 9.7b. Each set of metamorphoses is recursive. When we return to stage one we are witnessing a metamorphosis once more. This time it concerns the transformation of a spatial subsystem in region 1 (stage 4) into a point located in region 1 (stage 1 of the new run). We have to select at least one new class sharing the same space to start a new series of metamorphoses. The selection of classes is of course theoretically based, in non-geographical theories about relations between phenomena. However, since all of these classes are situated in the same space, which is the place in which all of these phenomena came into being, selection of classes should also depend on the characteristics of the place, or the issues which are considered to be important. In other words, our return to stage 1 not only implies theoretical necessity but also involves a regional question. The task is to relate the 'many spatial subsystems' of stage 4, although these are observed as points in stage 1 of the second run, to 'many new objects' in the new run.

It is not easy to train our eyes and brain in such a way that we perceive and consequently describe the unrolling landscape of changing forms. It is even more difficult to follow these metamorphoses when it concerns classes which are not sharing the same space, as is illustrated in Figure 9.7b. Superficially this figure resembles its partner. There are, however, three important differences. What happens during this second type of metamorphosis when spatially different classes all of a sudden form one superclass? First of all, we not only jumped from one level of analysis to another, from class to object, but we also entered another space, consisting of all of the spaces. In other words, the scale of the class, observed either via its members or as an object, remained the same in the first type of metamorphosis; in the second type we see that the scale of the class in stage 3 differs from the scale of the class that we meet in stage 2. Second, an extra metamorphosis takes place. Each spatial class which was selected in stage 1 is transformed into one object consisting of one or more areas of distribution (the filled spaces of the spatial class in stage 2). But its areas of distribution have to be changed into points (stage 3) in order to perceive all of these points, which are based on spatially different spatial classes, as one spatial superclass of N objects. Third, when we try a new run we again need more space, because any new class adds new space to the already selected spatial classes. Thus the ever-changing scales make the second type of geographical metamorphosis more complex to perceive than the first. Moreover, every change in scale depends on a decision about place because even scale, which is another word for space, is place. Our decision to extend our research on Swiss motorable passes to European Alpine passes must be based on a theoretical assumption which relates Swiss motorable passes to the others or to other categorically and spatially different classes. Moreover, we might have taken notice of a regional (i.e. Swiss) question which can be interpreted as a problem relating Switzerland to Europe. Subsequently, we must try to translate this regional link into a theoretical relation between spatial classes which occur in Switzerland and in Europe.

We started this discussion with the example of Swiss passes. Our first-selected 'space' was therefore Switzerland. While studying the distribution of these passes, two types of topical 'spaces', or thematic regions, emerged — areas of distribution and areas in which (parts of) areas of distribution were overlapping (the areal associations, all parts of Switzerland). Our second-selected 'space', the fourth to be introduced as an element in a scientific study of distribution patterns, was the Alpine region, or any other part of the world including Switzerland. The fifth space was part of this fourth space in the form of an area of distribution. The two selected 'spaces' are real 'places'. Even when we pursue spatial facts in order to transform them into general knowledge, we need at certain points in our analysis a 'place' to settle them: a place for which all of these spatial facts are an attribute or property. In other words, at certain stages in any analysis of spatial phenomena, 'space' becomes 'place'.

Geographical regions: the domain of regional-geographical knowledge and regional-geographical concepts

Spatial facts fall into 'places', whether we like it or not. As soon as we accept 'space' as 'place', our task as regional geographers is to let regional facts emerge out of the spatial information that was collected in the place which a while ago was considered 'space'. What difference does it make to see Switzerland as the 'place' of Swiss motorable passes and cantonal capitals instead of their 'space'? To answer this question we can use various maps: those on which either areas of distribution are indicated (Figures 9.1c, 9.2c, 9.1e, 9.2e) or areas of areal associations (Figures 9.4a, 9.4b). On all of these maps, our 'place' Switzerland is partitioned in different ways. In one respect they resemble each other, because they divide this place into regions of 'haves' and 'have nots'. From the spatial point of view, these empty areas are without meaning; from the regional point of view, they are as important as any other 'haves' region. They are all members of one class, i.e. the geographical region Switzerland.

What do we mean by a geographical region? It is a region consisting of regions, just as a soccer ball consists of hexagons and pentagons. As such, it is a complete class consisting of members having 'areal' and 'situational' attributes. We might say that we are able to interpret a class of hexagons and pentagons as a ball, owing to their situational characteristics, and as a soccer ball, owing to their shape and colour, or the areal characteristics. The same applies to a geographical region. We are able to perceive a place as a geographical region when the areal and situational attributes of a certain number of thematic regions are arranged according to some rules. When we decide to observe situational attributes of these regions as one attribute of whatever geographical region, we are trying to describe the areal integration of that geographical region. Simultaneously, we must try to describe the areal attributes of the thematic regions as an areal differentiation, which is the second attribute of a geographical region. In other words, any description of areal integration and differentiation represents a metamorphosis following stage 2, 3, or 4 in the series of metamorphoses given in Figure 9.7a. However, we call a certain type of ball a soccer ball and not a tennis ball, owing to the fact that the areal and situational attributes of the constituent parts of soccer balls and tennis balls differ. Geographers must have terms at their disposal with which they cover the theoretical relations between areal and situational attributes on a subregional level. Once they have these terms, they are able to formulate rules concerning the variation in areal and situational attributes. Applying the rules, they can design regional models. Any geographical region showing a pattern of areal differentiation and integration which reminds us of one of these models can be described in terms of the model and explained in terms of the rules. These concepts refer not to frequencies of spatial classes and their members in either a thematic or a geographical region but to frequencies of thematic or geographical regions in a geographical super-region. We must also realize that all types of regions, such as single- or multi-feature regions, administrative subdivisions of a country, provinces, or whatever regions are known by their toponyms, or in our mind, are thematic regions when treated as undivided units, and geographical regions when treated as classes of regions. Thus, the discussion among geographers should not concern the question of whether or not regions exist; the real issue at stake is the question of why we consider a certain 'space' as a thematic region and not as a geographical region, or vice versa. When we choose to treat 'space' in a geographical manner, we have to face the problem of how to relate the most fundamental geographical concepts of areal integration and areal differentiation to a preferably simple theoretical regional model. To solve this problem we should first establish which and how concepts referring to class attributes of thematic regions, which are observed as class members of a geographical region, are related to the concepts of areal differentiation and areal integration of a geographical region, observed as an object.

Thematic regions sharing one place: a geographical region

Attributes which are relevant for thematic regions as parts of a geographical region are given in Figure 9.8. These include: (i) their three-dimensional characteristics of shape, size, relief, and physical foundation; (ii) their time characteristics: period of observation (in historical terms), time-distance aspect in relation to period of observation, and the size of the region (cf. Figure 9.3); (iii) their categorial characteristics such as frequencies with which spatial classes and their members occur in each and every thematic region; (iv) the type of network and interaction patterns, or part of them, between the members of all of those spatial classes, selected for theoretical reasons which happen to be present in a thematic region; (v) the intensity and quality of intraregional traffic of immaterial and material products which originate in the spatial classes in a thematic region; (vi) the type of spatial system, or part of it, which is formed by the members of spatial classes in a thematic region.

The first set of attributes relates a thematic region, which is observed as part of a geographical region, to its physical domain; the second set relates it to a particular cultural-historical domain; the third, fourth and fifth attributes are the very basis upon which the area is defined and delineated as a thematic region. In case an area in a geographical region lacks one or all of these attributes, it nevertheless belongs to the class of thematic regions as a member of a geographical region, because absence is only the negative aspect of otherwise positive attributes. Depending on the shape and size of a thematic region, it receives a boundary with a certain length. Depending on its relief, it obtains a boundary of a certain quality, especially with regard to accessibility. According to the frequencies and quality of the perforations in the boundary — which are to be studied as members of a spatial class — the boundaries of some thematic regions are more open to others than those of other thematic regions in a geographical

Figure 9.8 Interrelated concepts from the domain of nature, history, space, and place on one level of analysis and applicable to a thematic region in a geographical region

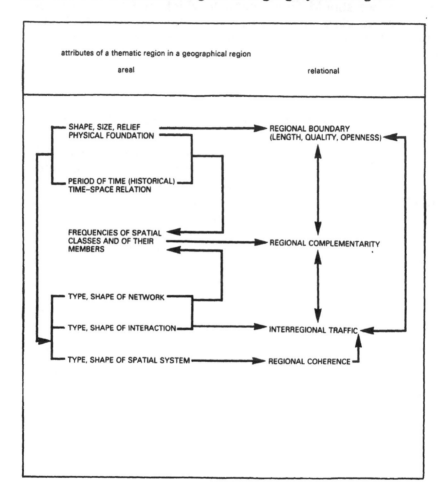

region. In other words, a thematic region can be characterized by variable degrees of openness, depending, among other things, on the length and quality of its boundary. Therefore 'regional boundary' is the seventh, rather comprehensive, attribute of a thematic region which is relevant for the geographical region. Interregional traffic is the eighth important attribute. It refers to those parts of networks and interaction patterns between members of spatial classes which fall outside the boundaries of the thematic regions that were established in the previous stage of analysis. For obvious reasons, there is a relation between the degree of openness and the possibilities for interregional traffic. Problems related to the size, shape, and relief of a thematic region create new time-space solutions, which sometimes are the introduction of a new period in history. Alternatively, time-space solutions might condition the shape and size of a thematic region. To a large extent these attributes together determine the frequencies, networks, and interaction patterns, which are the spatial attributes of the spatial classes occurring in the space, now observed as a thematic region. In turn they influence interregional traffic and, consequently, length and openness of regional boundaries. We assume that difference in frequencies between thematic regions is either the generator or the result of traffic between the thematic regions. We might as well cover this ninth attribute under the name regional complementarity. Regional complementarity and interregional traffic are interdependent. The more complementary thematic regions are, the more traffic they could generate; also, the more interregional traffic between the thematic regions, the more, or the less(!), complementary they become. In a positive way complementarity, which is always in combination with interregional traffic, is responsible for the tightening of the connections between thematic regions and, consequently, for the areal integration of a geographical region. In a negative way it stimulates separation between thematic regions and the areal disintegration of a geographical region (see Figure 9.9).

In case we discover regional complementarity, or interpret the variation in areal attributes as such, this implies that there are or must be interactions and connections between members of spatial classes in the corresponding thematic regions of the geographical region. This happens when the various networks and interaction patterns that were established in the previous round of our analysis (i.e. of the spatial classes involved, see Figures 9.3 and 9.5) are interlinked and mapped as the spatial subsystems (regions I-VI in Figure 9.10). Combined with the areal associations in Figure 9.4a and 9.4b, the six spatial subsystems become partitioned and, consequently, become geographical regions themselves. For instance, in Figure 9.10 we can see that the large subsystem, in the south-eastern part of Switzerland, is divided into an eastern region, a central region containing the valley and the capital, a western region, and two tiny regions along the western border where two thematic regions overlap. These are the Vorderrhine region with the Oberalp and Lukmanier passes, and the Hinterrhine region with the San Bérnardino and the Splügen passes. We are witnessing a change in level of analysis. A thematic region, which is a member of the geographical region Switzerland,

Figure 9.9 Conceptual framework concerning a geographical region which is observed as class and as an object

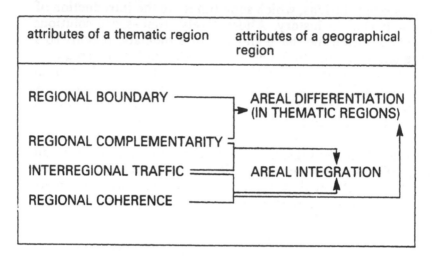

becomes the geographical Graubünden on another level of analysis. Simultaneously, Switzerland becomes more areally differentiated.

When we decide to reserve the concept of regional coherence, the tenth relevant attribute of a thematic region, for the extent to which thematic regions are interlinked through members of their respective spatial classes, we can formulate the following three rules. First, the more regional coherence there is, the more interregional traffic there is (see also Figure 9.8). Second, the more regional coherence there is between thematic regions in a geographical region, the more areally differentiated is that geographical region in other aspects (mostly in terms of spatial subsystems). Third, the more regional coherence there is, the more geographical regions the original geographical region contains. This third rule implies that a geographical region consists not only of thematic regions but also of geographical regions and that this new class is related to the thematic regions. Regional coherence is also responsible for a relatively strong areal integration on the level of subgeographical regions. It might therefore be a negative force on the level of the geographical region. We can formulate the following rule: the more regionally coherent that some thematic regions are in relation to others in the same geographical region, the less areally integrated that geographical region could be (see also Figure 9.9). The situational attributes of the subregions of a geographical region are: (a) the boundaries (type and length) that they have in common with each other, as well as the type and length of the boundaries that they share with the border of the geographical region; (b) their position with regard to each other and to the border; (c) the number and quality of the possibilities through which traffic can directly and indirectly enter regions that are situated inside and immediately outside the geographical region (for instance, via motorable passes). We can classify any type of sub-region in a geographical region, according to these situational attributes, as (a) border regions, the boundaries of which partially coincide with the border of the geographical region; (b) central regions, which are surrounded by other subregions; (c) intermediate regions, which are surrounded by central and border regions.

The frequency of border, intermediate, and central regions is the attribute of a geographical region which is observed as class (see Figure 9.11). There is still a second relevant attribute of a geographical region which is observed as class: the number of homogeneous, thematic regions that can be delineated on the basis of distribution patterns of one or more spatial classes which are present in the geographical region, and the number of heterogeneous, thematic regions that are based on the number of areas of areal association of these spatial classes which are observed as one spatial superclass. In case all, that is, one or more, spatial classes are equally distributed over their space (i.e. the potential geographical region), we find one single, or multi-feature, thematic region (or one homogeneous or heterogeneous thematic region respectively), instead of the expected geographical region that is divided into two or more thematic regions. So we should look for other spatial classes. Suppose one or more selected spatial classes are concentrated. We then are dealing with two thematic regions: one homogeneous or heterogeneous region and one which is empty (a special kind of homogeneity). These two thematic regions are complementary. Together they form one geographical region with the most simple areal differentiation, consisting of one core-like and one periphery-like thematic region. Of course the most likely situation is a geographical region consisting of thematic regions with different concentrations of members of spatial classes in different combinations. This implies that there are heterogeneous regions which are core-like, having a concentration of spatial classes and a large number of their members, and more

Figure 9.10 Areal associations of Swiss cantonal capitals and motorable passes observed as a spatial system (based on Figures 9.4a, 9.4b, and 9.1e)

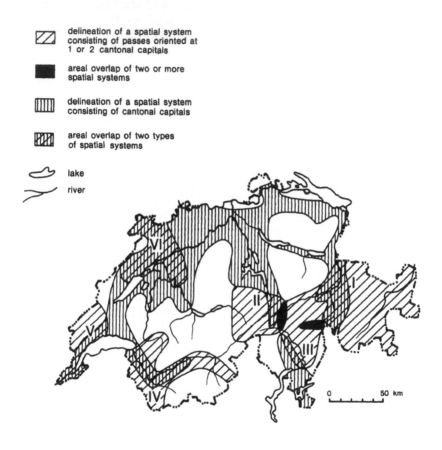

or less homogeneous regions which are peripheral in comparison with the others, containing only one or two spatial classes and relatively few of their members. Since the distinction between core and periphery does not illuminate the richness of the concept of areal differentiation, we may well add transitional types of thematic regions which are characterized by various degrees of concentration and heterogeneity. Whether we are dealing with a core, transitional, or peripheral region, we are confronted with thematic containers. However, the container concept is applicable only on the lowest level of regional analysis, since the geographical region is understood

Figure 9.11 Interrelated concepts on three levels of analysis and applicable to a geographical region

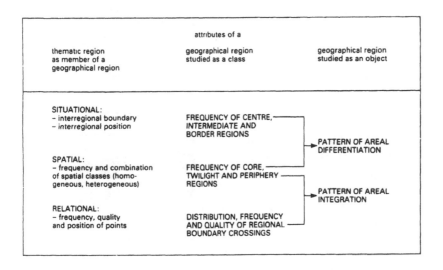

only via the relations between thematic containers.

When we describe the position of each core, transitional, and peripheral region in terms of centre, border, and intermediate region, we confront the reader with the pattern of the areal differentiation as one of the main attributes of a geographical region which is observed as an object (see also Figure 9.11). Since core regions are always complementary to transitional and peripheral regions, and regions can be qualified as centre regions through the regional boundaries that they share with other regions, we might say that the attributes of regional complementarity and regional boundary determine the pattern of areal differentiation (Figure 9.9). The frequencies, quality, and positions of points where interregional traffic can cross the border of a thematical region are attributes of these regions. The average number and quality per thematic region is an attribute of the class. Through this attribute, Switzerland, which is divided into thematic regions, can be compared to any other geographical region.

However, the total frequencies and distribution of the points where traffic can cross the interregional boundaries is

Figure 9.12 Four examples of areal differentiation of a geographical region

(a) Region with one core region in its centre region, twilight regions in its intermediate regions, and peripheral regions in its border regions

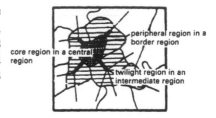

(b) Region with a peripheral region in its centre region, twilight regions in its intermediate regions, and core regions in its border regions

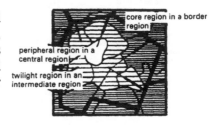

(c) Region with two cores, one in its centre region and one in an intermediate region, and twilight and peripheral regions in its intermediate and border regions

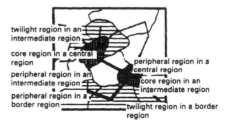

(d) Region with two core regions in border regions, a twilight region in its centre region, and other twilight and peripheral regions in the intermediate and border regions

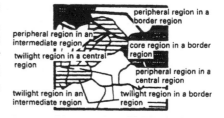

an attribute of a geographical region which is observed as an object. Together with the number and position of perforations in the border of a geographical region, we can interpret these points as an indication of or the result of the interregional traffic which determines to a large extent the areal integration and, together with regional complementarity, the areal differentiation of the geographical region. In case there are more points where important roads are crossing the border of the geographical region than points where important roads are crossing all regional boundaries, the geographical region might

be characterized by a tendency to disintegration. The history
of Switzerland is to a large extent the history of keeping
together thematic regions. Through this, its individual cantons
came into being as regionally coherent units, because all
mountain cantons can be delineated on the basis of pass
locations (Uri, Wallis, Graubünden, and Neuchâtel). Some
cantons became areally integrated or disintegrated through the
development of their passes in relation to interregional and
extra-regional traffic. Uri is perhaps the most dramatic
example. This canton was once the most integrated region in
the collection of cantons around the Vierwaldstättersee, Wallis,
and the region of the famous cloister Disentis in the eastern-
most part of Graubünden. The people of Uri (the Urner)
developed their passes in western, southern, and eastern
directions. However, this canton in the very heart of
Switzerland — historically as well as geographically — became
areally disintegrated in relation to Switzerland, following the
opening of the St Gotthard pass to modern traffic in this
century. This example may make it sufficiently clear that
analysis of the spatial class of points where roads are crossing
boundaries of thematic regions and the border of the geographical region must be incorporated in any study of geographical
regions. Otherwise the relation between interregional traffic
and areal integration cannot properly be understood. However,
we can only add this class when we have already delineated
the thematic regions on the basis of distribution patterns of
other spatial classes. Only when we treat the official divisions
of a region, say cantons in Switzerland, as thematic regions,
may we include this class of points right from the beginning.

Thematic regions in different places: models of geographical regions

Having discussed and related all of the concepts that refer to
attributes of regions and geographical regions containing these
regions, we now proceed to formulate the theoretical relations
between areal differentiation and areal integration and to
describe the corresponding models of geographical regions (see
Figure 9.12). The positions of the core regions are either in the
centre of a geographical region, or somewhere along its
border, or somewhere in between. We might propose the
theory that a geographical region is characterized by core
regions in its centre in case its spatial classes are not related,
or cannot be related, to spatial classes in which spaces do not
coincide with the geographical region. Consequently, the

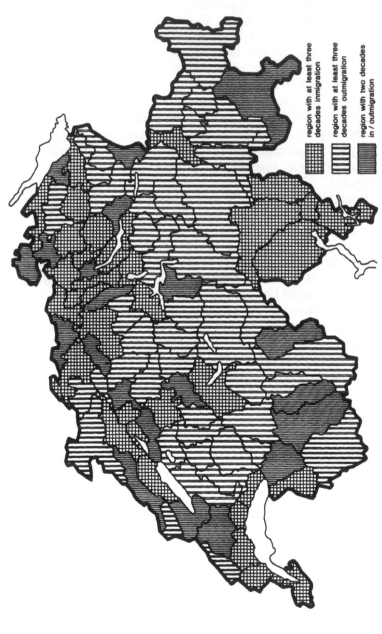

Figure 9.13 Areal differentiation of Switzerland based on migration patterns in subcantons between 1940 and 1980

region with at least three decades immigration

region with at least three decades outmigration

region with two decades in / outmigration

Figure 9.14 Five examples of matrices containing areal and situational attributes of all thematic regions of a geographical region and their frequencies

region	central	intermediate	border
core	1		
twilight		5	
peripheral			8

(a) Corresponding with figure 9.12a

region	central	intermediate	border
core			8
twilight		5	
peripheral	1		

(b) Corresponding with figure 9.12b

region	central	intermediate	border
core	1	1	
twilight		2	3
peripheral	1	1	5

(c) Corresponding with figure 9.12c

region	central	intermediate	border
core			2
twilight	1	2	2
peripheral	1	2	4

(d) Corresponding with figure 9.12d

region	central	intermediate	border
core	14	9	18
twilight	7	2	9
peripheral	17	17	11

(e) Corresponding with figure 9.13

Figure 9.15 Five examples of regional models or types of geographical regions based on a relation between areal and situational attributes of all thematic regions of a geographical region (Figures 9.15a-19.15e correspond with Figures 9.14a-9.14e)

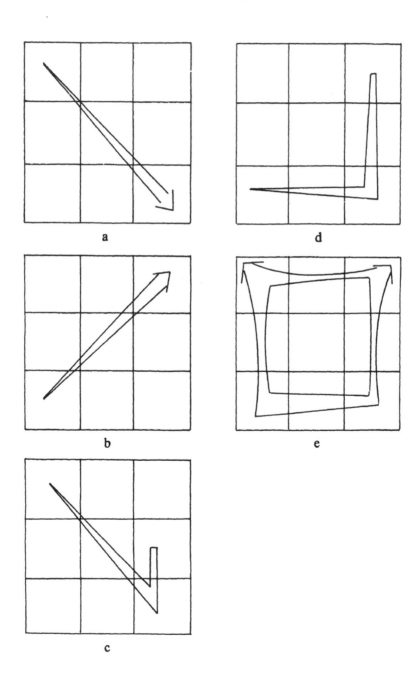

spatial classes of the geographical region, observed as objects, do not have external exchange relations with other spatial classes, which are situated outside this geographical region (see Figure 9.6), and the border of the geographical region is not perforated by passes, or international highways. A geographical region is characterized by core regions along its border when the spatial classes in the geographical region can also be studied as members of a spatial superclass in which space includes the geographical region. This region has a perforated border where the core regions are situated. These classes have external exchange relations with other spatial classes and are therefore best served by open borders and a short distance to the points where traffic must pass the border (Figure 9.12b). The areal differentiation of Switzerland on the basis of migration patterns during the last four decades of this century, which is given in Figure 9.13, cannot be characterized by any of these models and so we need another one which is flexible enough to describe adequately the endless variation of areally differentiated geographical regions. We propose a rather simple solution: a matrix which relates core, transitional, and peripheral regions to their situations in the centre, intermediate, and border regions. Figures 9.12a and 9.12b are maps showing the areal differentiation of the two ideal geographical regions with either a core in the centre and the periphery at its border, or a periphery in its centre and cores at its border. Figures 9.14a and 9.14b are the corresponding matrices relating the areal to the situational characteristics of the thematic regions of these ideal geographical regions. Figures 9.15a and 9.15b are transformations of the content of the matrices, showing the dominant relation between some areal characteristics and specific situational characteristics. Some geographical regions with more realistic areal differentiations are shown in Figures 9.12c and 9.12d. (Their corresponding matrices and types are given in Figures 9.14c and 9.14d and in Figures 9.15c and 9.15d respectively.) Thus, a geographical region can be classified on the basis of the thickness, direction, and number of the arrows in a nine-cell matrix. The same matrix can be used for data on the average number and quality of roads crossing the boundaries of thematic regions and the border that these regions share with the geographical region. As such, it provides insight into the relation between interregional traffic and areal integration. We take the view that, in a truly harmoniously developed geographical region, the core regions have the highest scores and the peripheral regions the lowest. This means that the matrices of areal differentiation and integration are almost identical.

Deviations from the expected pattern might be interpreted as signs of recent changes in the areal differentiation, or as remnants of old patterns, or as the result of recent or past changes in the shape and size of the geographical region itself.

In order to give an example of how a rather complex picture of areal differentiation might be reduced to a type, without losing the information of its complexity, we have analysed migration data of the last four decades for all subcantons of Switzerland (see Figures 9.13, 9.14e, and 9.15e). Regions with immigration during at least three decades are considered as core regions, subcantons with at least three decades of emigration as peripheral regions. Transitional regions have known two decades of emigration. On the basis of this map, Switzerland might be characterized as a geographical region with a core-like border, with a centre in which core and periphery are almost equally represented, and with an undeveloped intermediate zone. According to this pattern, Switzerland might be an example of a country where two types of geographical regions overlap: one based on new interaction patterns and one based on a traditional centre-oriented system. These new interaction patterns are served by international and national highways. The old interaction pattern is partly saved owing to decisions to adapt many old passes to modern traffic.

Conclusion

This chapter was written as a contribution to the methodology of regional geography. Geography is about regions. Therefore, regional geography is about regions. If this is not a tautology, what is the difference between regional geography and other geographies?

Geography is about generalities which describe and determine relations between attributes (or concepts) of classes. These classes are 'people' and land in which 'people' are actors, organizations, or institutions and 'land' can be considered as organized spaces or regions. In our view, any geography can be distinguished by its study of the type of organization or process that turns space into an organized space or a *thematic region*: physical geography is about the natural forces that form natural regions, political geography is about a combination of forces that form political regions, urban geography is about a combination of forces that form settlements of different types, and so on. Regional geography is about a complex of forces that binds natural, political, urban, and other thematic regions into more or less coherent regional units, or *geographical regions*, and those forces that unbind geographical regions into their constituent parts. Thus, there is both a substantial and methodological difference between geographical regions and thematic regions.

The substantial difference

The most distinctive attribute of a geographical region is that it consists of *classes of regions* which are contiguous, whereas the most distinctive attribute of a thematic region is that it consists of *classes of localized phenomena which are treated as if they were point locations*. Both geographical and thematic regions can be studied as classes which are distributed in a larger region. By doing so, we are easily eliminating their areal dimension. This is the reason why even large, complex geographical regions, such as towns or national parks, can be mapped and treated as points. In order to save their regional geographical attribute or differentiation pattern, they should be described and mapped according to a *typology of areal differentiation patterns*, in the same way in which we save the areal dimension of a thematic region by translating frequencies of localized phenomena in densities.

The methodological difference

The difference between regional geography and other geographical subdisciplines is also a methodological question. In regional geography, regions are studied on at least *two levels of analysis*. On the lower level, we can decide to study thematic regions as contiguous subregions of a geographical region. On the higher level, we study a region as a geographical region by analysing its type of areal differentiation as the outcome of our analysis on the lower level. In other words, in regional geography, thematic regions are studied as a class in order to transform them into an attribute of a geographical region. In other geographies, regions are studied on only one level of analysis: each is a member of a thematic class.

The subject matter of regional geography: geographical region as a concept of a class of phenomena

In the last part of this chapter we provided a model of areal differentiation. It is based on space-organizing forces and reflects the intricate relationships on various levels of analysis between spatial, thematic, and regional geographical classes and their attributes. The variants which can be deduced from this model can be used to classify geographical regions. Therefore, regional geographers do not need to bother about the general or scientific quality of their products. General knowledge is about classes of phenomena. Each geographical region is a class in itself. Each regional monograph informs the reader about general knowledge on subregional level, provided that all subregions are studied on the basis of the *same attributes*.

The subjective matter of regional geography: space or place?

Space, organized by people in and during a certain period of history, becomes place. Place, observed as a potential area of action, by either scientists or other organizers, becomes space, once again to be organized by them or other people, and so on. Regional geographers should describe the ever-changing scales and levels of analysis which people use to organize space into place. Simultaneously, they should be well aware that they resemble the people in a fascinating drawing by the famous artist Escher. It shows a building with people who are busy carrying objects to ever-higher levels but who nevertheless remain on the same level in the two-dimensional space that keeps them going. Escher called this human condition Relativity.

References

Allemann, F.R. (1985) *26 mal die Schweiz*, Munich: Piper.

Atlas der Schweiz (1965-1978), Wabern-Bern: Verlag der Eidgenössischen Landestopographie.

Blaut, J.M. (1962) 'Object and relationship', *The Professional Geographer*, 14(b):1-7.

Braudel, J.M. (1984) *Civilization and Capitalism, 15th-18th Century. III The Perspective of the World*. Collins, London.

Brugger, E.A. (ed.) (1984) *The Transformation of the Swiss Mountain Regions*, Bern: Verlag Paul Hapt.

Entrikin, J.N. (1981) 'Philosophical Issues in the Scientific Study of Regions', pp.1-27 in D.T. Herbert and R.J. Johnston (eds) *Geography and the Urban Environment*, vol. 4, Chichester: John Wiley.

Ferras, R. (1986) 'Ecrire de la Géographie Régionale sur l'Espagne', *L'Espace Géographique* 15:283-8.

Grigg, D.B. (1967) 'Regions, Models and Classes', pp.461-501 in R.J. Chorley and P. Haggett (eds) *Models in Geography*, London: Methuen.

Hall, R.B. (1935) 'The Geographic Region: A Résumé', *Annals of the Association of American Geographers* 25:122-36.

Hart, J.F. (1982) 'The Highest Form of the Geographer's Art', *Annals of the Association of American Geographers*, 72:1-29.

Johnston, R.J. (1984) 'The World is Our Oyster', *Transactions, Institute of British Geographers*. NS9:433-59.

Lukermann, F. (1964) 'Geography as a Formal Intellectual Discipline and the Way in Which it Contributes to Human Knowledge', *The Canadian Geographer* 8:167-72.

Piveteau, J.L. (1986) 'Identifier et Relativiser les Territoires', *L'Espace Geographique* 15:265-71.

Sack, R.D. (1974) 'Chorology and Spatial Analysis', *Annals of the Association of American Geographers* 64:439-52.

Scholer, M. and Bopp, M. (eds) (1986) *Strukturatlas der Schweiz*. Zurich: Ex Libris Verlag.

Whittlesey, D. (1964) 'The Regional Concept and the Regional Method', *Annals of the Association of American Geographers* 54:21-67.

10 Doing regional geography in a global system: the new international financial system, the City of London, and the South East of England, 1984-7

Nigel Thrift

Introduction: the internationalization of everything?

The world is becoming a more interconnected set of places as the internationalization of economies, societies, and cultures goes on apace. As Clifford puts it:

> This century has seen a drastic expansion of mobility, including tourism, migrant labour, immigration, urban sprawl. More and more people 'dwell' with the help of mass transit, automobiles, air planes. In cities on six continents foreign populations have come to stay — mixing in but often in partial, specific fashions. The 'exotic' is uncannily close. Conversely, there seem no distant places left on the planet where the presence of 'modern' products, media, and power cannot be felt. An older topography and experience of travel is exploded. One no longer leaves home confident of tackling something radically new, another time or space. Difference is encountered in the adjoining neighbourhood, the familiar turns up at the end of the earth.
>
> (Clifford 1988:13)

This is not to say, however, that regional diversity is dying out. Local difference can still be important, indeed may even have become more important (Lash and Urry 1987), if in different ways. For example:

> The people in my favourite town drink Coca Cola, but they drink *burukutu* too, and they can watch *Charlie's Angels* as well as the Hausa drummers on the television sets which spread rapidly as soon as electricity has arrived. My sense is that the world system, rather than creating massive cultural homogeneity on a global scale, is replacing one diversity with another; and the new diversity is based

relatively more on interrelations and less on autonomy.
(Hannerz, cited in Clifford 1988:17)

This chapter provides one example of how regional geography can face up to both the new global and local interrelationships of the world by considering the regional effects of perhaps the most international of all the international systems for extending and manipulating social power, namely the financial system. The chapter will be concerned to show how, in a specific time period, this system reproduced and even reinforced older patterns of socio-spatial differentiation of power, even as it was going through a period of accelerated internationalization.

Accordingly, the chapter is in three main sections. The first is concerned with articulating some of the major changes in the international financial system in the 1980s, up until the 'Big Crash' of October 1987, and the effects that these changes had on one of the chief nodes of the international financial system, the City of London. The second major section considers how these changes reached into the City in the period 1984-7 (as the so-called 'Big Bang'), and how they affected its labour market, especially by making it more international and more remunerative. In particular, I am concerned to argue that the boost to City incomes led to considerable accumulation of wealth, with important implications for the direction of national and regional class formation.

This argument leads to the third major section which considers how the City triptych of labour market change, wealth accumulation, and class formation can be translated into the arena of the South Eastern region of England. In particular, a direct connection will be made between this triptych and the pre-eminence of the South East within the British space economy via the demand for goods and services.

What can be seen at work through this case study of global and regional articulation is not a global financial structure 'out there', as a *deus ex machina*, but rather a spatially distributed network of money/social power which encompasses the globe. In this increasingly decentred network, local actions can quickly become global perturbations in ways unthought of when the international financial system was a more hierarchical, more centralized, and no doubt more elegant affair. Now the local and the global intermesh, running into and out of one another in all manner of ways. But this is not to argue that a hierarchy no longer exists within this network. The 'golden triangle' of New York, Tokyo, and the City of London still tells the fortunes of the world.

NIFS, state action, Big Bang

Crisis and change

Since the early 1980s, the world's financial markets have undergone a process of restructuring which has resulted in the emergence of a New International Financial System (NIFS; Thrift and Leyshon 1987). This new phase in the development of the international financial system has been characterized by a radical change in the nature and direction of international capital flows. It is also characterized by a further erosion of Pax Americana and the continued progression of Japan towards world economic leadership (Nomura Research Institute 1986).

The emergence of the NIFS marked a second major phase in the restructuring of the world's financial markets after World War II. The first phase of this process hinged upon the decline of the value of the US dollar in the late 1960s and the emergence of differential national rates of inflation. This led directly to the abandonment of the dollar's convertibility into gold in 1971 and also to the subsequent dismantling of the Bretton-Woods regime, which ushered in a period of floating exchange rates and volatility in the world's financial markets. The catalyst for the second phase of restructuring and for the formation of the NIFS was the developing country debt crisis of the late 1970s and early 1980s which dramatically altered the provision of international debt, and had far-reaching consequences for borrowers and financial institutions alike.

From debt crisis to closed financial system

At the heart of the developing country debt crisis was the recycling of the 'petrodollars' of the oil producing states into loans for third world sovereign borrowers. Following the 1973 price rise by the OPEC nations, the international banks performed intermediary roles in the redistribution of the funds that flowed from the oil-producing states to the Euromarkets that were based in London. The international banks banded together in lending syndicates to redistribute the capital to third world sovereign borrowers. By 1979 the volume of syndicated loans, which were arranged in the Euromarkets, reached $80 billion (Plender and Wallace 1985).

The organization of debt in this way fell apart in the early 1980s however, hitting both borrowers and intermediaries. Developing debtor countries were severely affected by the worldwide deflationary trends that followed the oil price fall from 1981 onwards, which brought in its wake high real interest rates and damaging new repayment obligations. Furthermore, the monetarist policy of the Reagan administration and its expansion of the US budget deficit simultaneously pushed interest rates higher and strengthened the dollar, which further increased the real repayment costs of dollar-denominated debt. Consequently, borrowers started to suspend repayments in 1982. By 1984 some thirty-five sovereign borrowers were unable to service their debt (Plender and Wallace 1985).

The debt crisis abruptly ended the recycling of capital from the first to the third world and created a closed financial system where funds are transferred predominantly between western industrialized nations. This restriction of capital flows resulted from the behaviour of the three principal groups that were involved in the organization of debt: bankers, borrowers, and investors. First, to prevent any further entanglement within the crisis, banks stopped lending money to third world sovereign borrowers by effectively drawing up a blacklist of states (Lim 1986). Second, in the wake of the crisis, many banks suffered a decline in their credit-worthiness ratings, so that western sovereign borrowers and multinational corporations discovered that it was they who were the better investment risk and that debt could be obtained on better and more flexible terms by issuing securities rather than by going through the intermediation of a bank. Third, this 'securitization' of debt was bolstered by investors who preferred not to deposit capital within the commercial and money centre banks that were directly embroiled in the debt crisis but instead to lend directly to borrowers via the purchase of their bonds and securities (Plender and Wallace 1985).

Investor power

Of the three groups mentioned above — banks, borrowers, and investors — it was the investors that were the driving force behind the NIFS. Pension funds and insurance companies, which had amassed large capital sums that needed to be invested to cover future disbursements, had developed into highly professional institutional investment vehicles. They were responsible for over two-thirds of the funding on most stock exchanges and exercised considerable customer power. The search for high-quality investments saw the institutions become increasingly international in their investment outlook.

Although all of the western industrial nations have their own institutional investors, by far the most important in the NIFS were and are the Japanese. In 1985 Japan became the world's leading creditor nation, and its investments in foreign securities totalled some $128 billion (Nomura Research Institute 1986). The outflow of money from Japan was further stimulated by the success of Japanese multinational corporations which sought investment outlets for their cash surpluses at home. The inability of the Japanese economy to absorb all of this money saw corporate profits and the funds of investment institutions increasingly forced overseas. The flow will inevitably increase still further in future as the progressive deregulation of the Japanese financial system goes on. Consequently, Japan's gross balance of overseas assets is expected to surpass $1 trillion by 1995 (Nomura Research Institute 1986).

The global centralization of financial service

In the past decade, the emergence of the NIFS has brought about a considerable realignment in the rankings of leading international banks. There have been two main changes. First, there has been a marked decline in the status of US money centre and commercial banks, such as Citicorp and Bank America, whose balance sheets grew fat during the 1970s on the loans that they arranged for third world sovereign borrowers, especially in South America. The leading banks of the NIFS are now Japanese. Their assets are buoyed by the massive foreign earnings of Japanese multinational corporations. Second, the move towards the securitization of debt has acted in favour of investment banks and securities houses. Prior to the advent of the NIFS, such institutions acted primarily as underwriters and traders of securitized financial instruments such as government securities and equity stock. The securitization of cross-border international debt has meant that the investment banks and securities houses were well placed to perform a similar role in the Eurobond market. The prerequisites for banking success in the NIFS are sufficient capital adequacy to be able to underwrite whatever size of bond issue that is required by a borrower plus an efficient distribution and sales network which is able to unload the bonds on to investors. These are characteristics more typical of securities houses and investment banks than the loan-based giants of the 1970s.

The transformation of international debt also meant an upheaval in the relationship between borrower and client banks. Long-term 'relationship' banking was largely replaced by service-driven links, a process which ultimately resulted in the centralization of financial provision among the world's leading banking institutions. The ability to provide an effective securities distribution system and a comprehensive set of financial instruments is an option which is open only to the largest and most powerful banks.

Technological change

Technological change played an important part in facilitating the development of the NIFS. Advances in information technology are constantly improving the external and internal efficiency of financial markets (Ayling 1986). The external efficiency of markets is improved by the increasing speed and capacity of financial information systems which enable market participants to make better informed and quicker decisions. Data production and distribution are increasingly subject to computer application and data are now supplied 'on-line' to market practitioners by financial information companies. The internal efficiency of markets is enhanced by improved links between markets, institutions, and individuals. Dealing and settlement systems are increasingly automated while electronic links enable traders to participate in 'remote' trading. In the NIFS, technological change is transforming securities trading from a primarily exchange-based activity into an electronic market place which is conducted within the dealing rooms of financial institutions.

The increasing automation of financial transactions dramatically increased turnover in financial markets. The number of transactions increased for three main reasons. First, the development of automatic-execution systems for the trading of small lots of securities speeded up dealing. Second, turnover was increased by the progressive introduction of 'expert systems' in financial trading which automatically alerted dealers to profit-making opportunities as they arose in the market. Third, the reduction in execution time and in costs, which was facilitated by technological application, encouraged institutional investors to become more 'active' in their management of investment portfolios, a tendency which was further encouraged by banks who were bidding for the right to restructure the entire portfolio of investment fund managers (Lee 1984).

New markets and capital substitution in the NIFS

The interaction of technological change and securitization
spawned a series of new financial markets as financial
instruments were increasingly transformed into tradeable
commodities. Relatively recent innovations, such as financial
futures and options, are now commonplace in the NIFS. There
is even a secondary market in third world sovereign debt
where developing country borrowing repayments can be
purchased at a discount of the original loan (French 1987).
However, perhaps the most important new market in the NIFS
was the emergent 'swap' market which brought about an
increasingly integrated world financial system. The burgeoning
swap market was a largely unregulated market which emerged
during the early 1980s. It comprises a series of financial
transactions in which two parties agree to exchange a pre-
determined series of payments over time (Hammond 1987).

The earliest swaps were currency based and were used to
arbitrate between different financial regulatory environments.
However, currency swaps have been surpassed in volume by
interest rate swaps which pivot upon the different credit
perceptions of borrowers between segmented markets. The
success of the swap is based upon its capacity to overcome
market imperfections. Swaps allow borrowers to engage in
transnational capital substitution by removing imposed market
variations. By allowing such substitution, swaps can be seen as
an important driving force in creating a more integrated global
market.

Spatial centralization of financial activity

Despite the fact that technological advance is increasing the
locational choices of financial activity as never before, a
process of spatial concentration is under way. In particular,
activity has concentrated in three major centres: New York,
Tokyo, and London. This process is a reversal of the trend of
the 1970s when there was a multiplication of financial centres
and a general lessening of the disparity between centres
(Browning 1986). The three major centres accounted for
almost 50 per cent of all international banking activity in 1985
and they were the largest capitalized stockmarkets in the
world. This concentration of activity arose, firstly, because of
the role that each centre performed as a regional financial
centre — New York for North America, Tokyo for Asia and
the Pacific, and London for Europe (and the Euromarkets) —

and, secondly, because of the strategic location of each centre within different time zones. The respective position of each centre enabled a continuous global market to develop whereby transactions could be pursued for 24 hours a day by passing on deals from market to market. For example, in a working day lasting from 6.00 a.m. to 6.00 p.m., communication and dealing technology enables a dealer based in London to catch the end of trading on the Tokyo exchange, to trade through the entire working hours of the London market, and to work for over half of the hours of the New York exchange. Although only foreign exchange dealing is a truly developed 24-hour global market at present, other markets, such as secondary trading in Eurobonds, international equities, and futures, are increasingly becoming so.

The impact of the NIFS on financial centres

For most of the post-war period, regulatory controls in financial centres were nationally based, instigated by governments who wanted to establish orderly domestic markets (Price 1986). The most common forms of regulation were controls of external capital flows (exchange controls), limits on interest rates, and the separation of markets and practitioners into clearly divided functions. The latter form of regulation purposely segments and partitions financial markets. For example, until October 1986, entirely separate roles were performed in the UK by commercial banks, merchant banks, stockbrokers, stockjobbers, and building societies.

However, changes in international markets, which were associated with the emergence of NIFS, made the operation of closely regulated domestic markets increasingly untenable. First, the broadening horizons of the investment institutions led to capital flows becoming increasingly mobile and international as funds sought investment outlets which were based on a risk/return allocational criterion rather than on an administrative rationale. Consequently, funds began to move towards less highly regulated financial markets. Second, the advent of securitization and disintermediation effectively fused the activities of debt provision with those of securities trading. However, the rigorous division of these activities, which was insisted upon in many domestic financial markets, meant that many institutions were hamstrung in their ability to adapt to the changes unfolding around them, since they were usually allowed to engage in banking or in securities trading but not in both. Finally, the ability of regulatory authorities to control the activities within financial markets was progressively undermined by the increasing application of new technology. Technological advances and volatility within financial markets gave birth to a plethora of new financial instruments which were beyond the control (and frequently beyond the comprehension) of the regulators. At the same time, transactions increasingly became electronically based, thus removing the need for fixed central market places, the traditional power-base of the regulatory authorities.

Faced with the combination of the internationalization of capital, securitization, and new technology, many financial centres realized that survival in the NIFS meant attracting both capital and international financial institutions: attempts to impose stringent controls were becoming less and less effective and therefore ultimately self-defeating in that they tended to drive away capital. As a result, the 1980s saw the worldwide dismantling of domestic financial restrictions. Australia, Canada, Finland, France, Norway, New Zealand, Portugal, Singapore, Sweden, Switzerland, South Korea, and West Germany were among those countries that embarked upon the progressive deregulation of their financial systems (A. Hamilton 1986; Ayling 1986). All of these moves were designed to help to capture a share of the burgeoning cross-border financing business in the NIFS. Similar moves were also made in the three premier world financial centres. There was New York's early abolition of fixed commissions in securities trading in 1975, and the establishment of 'offshore' International Banking Facilities in 1981 (A. Hamilton 1986). Tokyo too is in the grip of a (slow) process of deregulation which will allow Japanese capital greater access to world financial markets and foreign financial institutions greater access to the Japanese financial market (Marsh 1986). However, it was within the City of London that the adaptation to the conditions of the NIFS was most radical. State-imposed change forced the complete overhaul of practices within its domestic securities markets.

Deregulation in the City of London

The transformation of the world's financial markets in the wake of the NIFS meant that, during the late-1970s and the 1980s, the City was increasingly placed at a considerable competitive disadvantage. There were two main problems. First, the widescale application of technology in financial markets served to make the process of capital formation in the

UK look both expensive and inefficient because of its exchange-floor based, single-capacity, and fixed-commission system. Given the increased mobility of capital, this threatened the long-term attractiveness of London as an investment outlet. Second, in a period when financial centres worldwide had been loosening regulatory controls, the domestic UK financial system remained sharply balkanized. London's rise to prominence in the 1960s and 1970s was based on the regulation-free environment that it offered to international banks which were operating in the Euromarkets. However, when the emphasis in international debt shifted from loans to securities, UK institutions were stymied by their historical compartmentalization of function. They had difficulty in competing with the more diversified European 'universal banks' and with the investment banks and securities houses of the USA and Japan, which were better suited to the sales-driven orientation of the new securitized environment. The UK equivalent of the Japanese and American securities houses, the member firms of the Stock Exchange, remained comfortably cosseted, however, by the large earnings that could be made in the restricted, fixed-commission domestic securities market and they did not venture too far into the more competitive international arena. At the same time, the weakening pound and the size of the UK domestic market meant that the capitalization of the internationally oriented, British-owned institutions was significantly lower than that of their foreign competitors.

The restructuring of the City of London was a process that was initiated by state action in 1976 and was finally agreed to by the Stock Exchange in 1983. It was designed to overcome the two problems outlined above. The 'Big Bang', as it became known, essentially revolved around a state-enforced transformation of one financial market, the market for domestic securities. However, the restructuring of the equities and gilts markets was used as a catalyst for a wider reorganization of UK financial institutions, with the underlying hope that the synergy that was achieved from the fusion of formerly separate financial institutions would propel UK institutions to the forefront of the NIFS.

The 'new City' was unveiled on 27 October 1986. The totem of change was the scrapping of fixed commissions on all domestic securities transactions, a move strongly welcomed by the investment institutions in the equities market and by the state itself in the gilts market. However, this move in turn necessitated other changes. First, under a regime of competitive commissions, the single capacity system of market-making jobbers and agency brokers was considered to be unworkable. Previously, jobbers had earned income from the difference between the buying and selling prices of stock, while brokers had earned theirs from charges to clients for transactions that were executed in the market. It was felt that brokers would not be able to survive on reduced commission fees alone. Hence, a dual capacity trading system was introduced whereby individual firms would be able to perform both market-making and agency roles. This led, in turn, to a second necessary change: the opening up of the exclusively domestic members of the Stock Exchange to outside financial institutions. The restrictive nature of the fixed-commission, single-capacity system allowed firms to operate at remarkably low levels of capitalization. Indeed, all member firms had operated as partnerships.

However, in a competitive, dual capacity regime, greater levels of capital would be necessary to allow participation in larger and more frequent deals and to protect profit levels as margins declined. The necessary injections of capital came from UK and overseas financial institutions which, on the announcement of the opening up of the exchange, embarked on a frenzied shopping spree as they anxiously purchased expertise in domestic securities within the UK market.

The reformulation of the market was completed by the instigation of a screen-based dealing system. SEAQ (Stock Exchange Automated Quotations), which is closely modelled on the US over-the-counter electronic dealing system NASCDAQ (National Association of Securities Dealing Automatic Quotations), soon removed dealers from the floor of the exchange into the dealing rooms of the new financial conglomerates.

Impacts of deregulatory change in the city

The restructuring of the market had two major impacts upon the functioning of the City. First, the centralization of capital, which had already occurred as a result of the break-up of the balkanization of the market, increased still further. The overcapacity in the new market led to a shake-out of participants. Before the 'Big Bang', the equities market contained five main jobbers and three smaller ones; the new market contained 35 market-makers. Similarly, in the gilts market, the two largest jobbers had accounted for over 75 per cent of turnover; the new market contained 27 primary dealers (Farmborough 1987). The scale of overcapacity in the gilts market could be gauged by the fact that the US government securities market, although ten times the size of the UK market, supported only 37 primary dealers (Hewlett and Toporowski 1985). As a result of the increased competition, profits were slashed as equity transactions fell by as much as 50 per cent (Ingram 1987; Barrett 1987). Moreover, a small group of firms quickly rose to dominance, so that in the equities market the leading ten firms accounted for 80 per cent of all trading (*Economist* 1987a). The competition was already beginning to tell: Midland Montagu withdrew from equity market-making in March 1987, and Lloyds Bank pulled out of gilts three months later.

Second, although the British banks entered securities trading for the first time, thereby transforming themselves into entities that resemble the European universal banks, British purchases were more than matched by foreign buy-outs of stock exchange firms, which thus resulted in a sharp increase in the foreign presence in the City. At the time of the 'Big Bang', over half of the 27 firms making a market in gilts were overseas institutions.

To summarize, the City, faced with an increasingly inter-nationalized financial system, which was largely dominated by foreign institutions, had two options open to it. It could either remain firmly behind a barrier of restrictive controls, as business steadily migrated elsewhere, or it could attempt to introduce international 'financial best practice' (Goodhart 1987) into its operations. Forced by state pressure into the latter course, the viability of the City as a financial centre was considerably enhanced. However, it was at the cost of dimin-ished national involvement in domestic financial transactions.

Labour market, health, class

The changes that were induced by the forced adjustment of the City of London to the NIFS had numerous urban and regional effects. Many of these have been summarized elsewhere, including: the office building boom in the City of London and elsewhere; the effects on the rental and freehold housing markets in London; changes in the level and pattern of commuting in and out of London; multipliers from the expenditure on new technology in the shape of jobs in computing and telecommunications firms (Leyshon *et al.* 1987). The horizons of this chapter are limited to three interrelated effects, namely the City's labour market, the generation of wealth, and the changes in social class.

The City of London labour market

There has always been an international professional and managerial labour market but, for numerous reasons, it has of late been growing dramatically in scope. These reasons include: the increasing internationalization of corporations which are already multinational (especially as Japanese and European corporations have moved into the United States); the increased propensity to regard an international posting as a vital part of a career in a corporate bureaucracy; the increase in the number of international financial centres and of jobs in these centres. The case of the UK is distinctive in this regard. Findlay (1987), by using International Passenger Survey data, has shown that there has been a steady increase in the number of professional and managerial workers leaving and re-entering the UK in the 1980s, signifying, in the main, an increase in short-term job transfers.

A special subset of this professional and managerial labour market is in the international labour market for those working in financial services. Of late, this part of the international labour market has been growing particularly rapidly, in line with the growth in financial services worldwide (Key 1985). It is particularly orientated to the so-called world cities, such as London (see Thrift 1986; 1987c). However, until the period leading up to the 'Big Bang', from 1984 onwards, the City of London had been able to insulate itself to some extent from the international labour market (see Cobbett 1986).

The growth in foreign banks, which was coincident with the rise of the Eurodollar market in the late 1960s, was the first real crack in the clannish, clubbish wall of privilege and tradition that surrounded the labour market practices of the City. Aggressive New Yorkers were suddenly to be found in the confines of the City. The growth of foreign exchange trading, which followed the floating of exchange rates in the early 1970s, meant another crack in the wall. (The City had to meet the threat by allowing working-class clerical labour with sharp wits to move on to the dealing floors.) The growth of the Eurobond market in the late 1970s again brought more foreign banks and foreigners into the City. Each of these events meant not only an increase in the presence of foreign firms in the City, bringing with them labour practices which were foreign to the City, but also more Britons working abroad and being introduced to foreign labour practices whilst there. However, the events from 1984 onwards changed the City labour market irrevocably, fully articulating it into the international labour market and bringing other changes as well. From 1984 to 1987 the labour market underwent at least five significant changes. First, there was a growth in simple absolute terms, which was coincident with a general increase in employment in financial services and with a bull market. Exact numbers are not available but the 25.9 per cent increase in employees in foreign banks in 1986, from 42,767 to 53,833, is a particularly dramatic example of this growth. Second, as already inferred, the labour market became more international. There were more Americans, Japanese, Europeans, plus expatriate Britons returning to their homeland. The numbers were quite dramatic; one estimate is that, by 1987, there were 40-50,000 Japanese in London (Holberton 1987).

Third, the City labour market became more 'skilled'. Graduate employment in the City became something of a norm whereas previously it was relatively unusual. In 1985 just over 18 per cent of British graduates went into finance, many into the City (see Pagano 1986; O'Leary 1986). Fourth, and related to the third change, the City labour market became more specialized. As financial markets became more numerous and complex, and as methods of analysing their twists and turns became more sophisticated, so the demand for specialists increased, especially those with appropriate qualifications. For example, maths and natural science graduates became quite common.

Fifth, the labour market became much less rigid in terms of mobility. People became willing to switch jobs. They were looking no longer for a sinecure but for early responsibility and 'serious money'. The issue of 'serious money' leads to the next sub-section.

The generation of wealth

Wage rates in the City labour market were at one time cheap by international standards. However, during the 1980s, this situation began to change. Salaries began to catch up with those in other international financial centres. There were five main reasons for this narrowing of the wages gap. First, with the increase in the permeability of the City labour market to the international financial labour market, there was bound to be some levelling up of salaries. Second, substantial skill shortages became apparent as the new financial markets of the NIFS flickered into existence, as existing markets (e.g. dealers in Eurobonds) expanded, and as the rate of technological change hotted up (e.g. the need for software writers). This phenomenon was only helped by increased specialization.

Third, firms, especially large foreign firms, began to bid for this labour, often quite frantically: in so doing, they forced up the level of salaries. Fourth, information on the labour market became easier to find. As salaries became an important issue, so several specialist firms started up in order to supply information on the going rates for different kinds of employee. Finally, people working in the City themselves became more aware of their worth. Thus, as a general phenomenon, wages in the City of London moved upwards, in many cases substantially. Average salaries in the City diverged from those of the South East of England, and even more strongly from those of the UK as a whole.

The upward movement of salaries was distributed (unequally) amongst four main groups. First, there were the directors and partners of the various City firms — the so-called 'icing'. These people were already well paid before the 'Big Bang'. In 1985, 361 directors in 42 companies earned more than £100,000 per year; 24 of them earned more than £250,000 per year (*Labour Research* 1986). As the process of centralization took place, the overall financial position of these people was improved somewhat by the buying out of the directors and partners of many City firms. Some of them walked away with sums of £1 million or more.

Second, there were the top managers and dealers — the so-called 'marzipan' layer. These people began to earn very substantial salaries. In 1985, before the 'Big Bang', only 67 employees in 14 companies earned more than £100,000 (with nine earning more than £250,000). In the period leading up to the 'Big Bang' and just after, this position changed, especially if a person had skills which were in short supply. Salaries of over £100,000 became fairly common. One estimate is that, in August 1986, about 2,000 people, most of them drawn from the 'marzipan layer', were earning salaries of £100,000-plus (Pagano 1986). It is important to note here that many of the people in the 'marzipan layer' had their salaries boosted by various 'golden hellos' and 'golden handcuffs', often of up to £200,000 (*Guardian*, 12 August 1986). These were intended either to tempt employees to new firms or to keep them satisfied with their current firms.

Third, there were large numbers of people, especially but not exclusively young or youngish graduates, who were not, or not yet, in the high-flying positions. They all saw their salaries increase substantially, from those with starting salaries for graduates, averaging £16,000, right through to those higher up the career ladder but not yet at the top. Differentials were often to do with skill shortages. The example of computing experts is instructive in this regard. There was a general shortage of computing skills in the City and, as a result, computing jobs commanded, on average, a 10 per cent premium over other specialists (May 1987). Thus, compared with national figures, salaries were generally higher in the City for the same job, at a younger age.

In each of the above three categories, the figures mentioned were only salaries. They did not include other important monetary incentives. In particular, bonuses were vital. They brought some people's annual earnings to well above £100,000 overall and they generally upgraded the earnings of very many others. In addition, there were various perquisites, most notably a carefully graded company car, health insurance, life insurance, and mortgage subsidy, which again added to the overall earnings, even if indirectly.

One final category of workers in the City is yet to be mentioned: the lower paid, constituting probably about half of the workforce and concentrated in clerical jobs, personal services such as chauffeuring, and so on (Labour Research 1986). Many of these jobs were held by women. A secretary in the City could expect to earn £10-12,000, in addition to receiving a mortgage subsidy, cheap loans, health insurance, a pension scheme, and incentive payments, all of which could add another £2-3,000 in value (*Sunday Times*, 24 May 1987:49). This was quite high in comparison with national figures (but note the higher London cost of living) but was not high comparatively within the City, especially since secretarial work was becoming deskilled and repetitive (through the introduction of word processing systems). Further, salary figures like this did not apply to the many temporary clerical workers in the City, who were often paid considerably less (see Townsend 1987).

In general, however, the earnings explosion of the 1980s meant a vast increase in personal wealth for a considerable number of people. This City wealth can be quantified to a degree in three different categories: earnings, total income, and wealth.

Probably at least one-half of the City workforce (some 190,000 people) were earning more than the top 10 per cent (pretax) national wage of £16,681 in 1986. Of those, I estimate that about 4,000 people, including directors and partners, were earning, with bonuses, £100,000 or more in 1986. Some of course were earning substantially more than £100,000. With bonuses, a further 10-15,000 people were earning between £50,000 and £100,000 and other 80-100,000 were earning between £20,000 and £50,000. Using the most recent data on income distribution (Board of Inland Revenue 1987), it is possible to estimate that in 1986 those in the City had 11.5–14 per cent of all incomes in the UK between £20,000 and £50,000, 23–34 per cent of all incomes in the UK between £50,000 and £100,000, and 50 per cent of all salaries over £100,000 (see Thrift 1987).

In terms of total income (including income from rents, stocks and shares, dividends, bank and building society interest, etc.), these percentages might even increase. City people would naturally invest much of their money but they would be able to obtain a better return, through extra expertise and information. The same stricture applies in terms of property speculation.

Finally, in terms of wealth, City people would clearly have amassed considerable wealth. It is important to remember that a number of those working in the City already emanate from upper-class and upper middle-class backgrounds and are likely to be more wealthy than the population as a whole. Thus, to those that have shall more be given. Taking this factor into account, of the 210,000 millionaires to be found in the UK in 1986 (Shorrocks, cited in Rentoul 1987), probably at least 3,000 (or 15 per cent) were in the City. For those who are moderately wealthy (the top 0.1 per cent of the population with assets of over £740,000 [*New Society* 1986]), the figure is probably nearer 30 per cent.

Class

The changes in the City labour market, which are manifested in the growth in salaries, are important in understanding class restructuring in the City. Traditionally, the City has been a preserve of the British upper class, 'men who walk in lockstep from Eton to the grave, lingering in the City *en route*' (*Economist*, 1985, p.40). Even now the City is surprisingly homogeneous in class terms, at least in its upper reaches. *Becket's City Directory* reveals that one-quarter of top UK merchant bankers and stockbrokers went to Eton (Bowen 1986) and a public school background is the rule rather than the exception for those at the top of the occupational hierarchy. On average, some 74 per cent of top executives in merchant banks, stockbrokers, clearing banks, accountancy firms, insurance companies, and insurance brokers went to public school. A staggering 96 per cent of top stockbrokers had a public school background. This narrowness of educational background amongst those at the top of the executive hierarchy is reproduced at University level. On average, 68 per cent of the top City executives with degrees went to Oxford or Cambridge. This is no surprise. Until quite recently, many city institutions had a tradition of automatically interviewing all candidates from Oxbridge, while candidates from other universities were screened in advance of interview. These indices of upper-class supremacy at the top of the City ignore the presence of the aristocracy, which is extensive. For example, the merchant bank Kleinwort Benson had six lords amongst its employees, including some hereditary peers (*Business*, June 1987, p.18).

Beginning in the late 1960s, the upper-class ascendancy in the city did begin to break down, however. The first shock to the City was the influx of American banks in the 1960s in the Euromarket boom. These were more likely to recruit meritocratically and they trained people for early advancement. The floating of exchange rates in the early 1970s was another shock to the system. This ushered in 'the lads from Newham', or working-class foreign exchange dealers. Finally, the 'Big Bang' also opened up the City, for at least three reasons. First, recruitment procedures became more systematic and promotion procedures more meritocratic. Second, the restructured firms became larger, making it more difficult for them to recruit exclusively from the upper class or to keep an exclusively upper-class atmosphere. Third, and in a related vein, because of the wider competition for Oxbridge graduates, many firms were forced to take graduates from other universities. As a result of the 'Big Bang' there was, therefore, a general dilution of the upper-class presence in the City, but it is by no means certain that there was a lessening in absolute numbers. Indeed, given the greater flow of Oxbridge graduates into the City representation, this may have actually increased. For example, 26 per cent of Oxford's graduates went into commerce or chartered accountancy in 1986, many of them into the City (Eykyn 1987). The figures were somewhat less for Cambridge but in both cases they have been growing rapidly in recent years.

It is certainly too early to tell whether the influx from universities other than Oxford and Cambridge means that the social class profile of the city will be indelibly changed as these graduates work their way up the City career ladders, turning the City into a service-class preserve. There are certainly forces working against that possibility. First, the expansion of foreign banks and securities houses in the lead up to the Big Bang has not necessarily meant a bias against the old class ways. Indeed, foreign banks and securities houses took the second largest proportion of Oxbridge students. The major Japanese securities house, Nomura, took 28 graduate students in 1986, all from Oxford or Cambridge. Second, the social linkages of the 'old boys' network', although weakened by the influx of new players, still influenced employment opportunities. Even in 1986, an eminent stockbroker could still suggest that the best way to obtain a job interview was through 'the personal recommendation from some family friend', since it is 'much more difficult to refuse an interview if it is at the request of a mutual acquaintance' (J.D. Hamilton 1986:177). Third, much of the influx of graduates seemed to be a replacement for formerly 'unskilled' labour, such as dealers with working-class backgrounds. Fourth, it is difficult to know whether a filtering of graduates will not take place in the future in which only Oxbridge graduates become the top executives. Alternatively those middle-class graduates who succeed may simply be incorporated into the upper class.

What does seem to be happening is that it has become increasingly difficult for 'upper-class idiots' to survive in the restructured City of London. In Lloyds, for example, 'there are a number of (upper-class) brokers swanning around with not necessarily a lot between the ears' but if 'the families are still there, ... only the bright members survive in terms of becoming a big noise' (Bowen 1986:41). But it is a big step from observing the winnowing of the less mentally agile upper class to saying that the City will no longer be an upper-class preserve. If anything, it might be argued that the injection of wealth into the City before and after the 'Big Bang' enabled many upper-class people already in the City to consolidate their wealth and induced many more, brighter, young upper-class people to move on to the first rung of the earnings ladder than would have been the case previously.

The City in the South-East region

The South East of England has a long history of regional dominance in the UK. Recent work in economic and social history has helped to confirm this, demonstrating that much of the dominance of the South East in the nineteenth century was the result of a combination of 'international trade and particularly investment, the consumer demand sustained by many centuries' accumulation of landed wealth and the self-sustaining capacity of affluence' (Lee 1984:154). In other words, wealth, pure and simple, and the buoyant demand for goods and services that this wealth generated accounted for a good part of the relative dominance of the South East; 'an affluent, service-owned economy can to some considerable degree sustain its own well being' (Lee 1984:153). In turn, much of the fund of wealth of the South East was accumulated (and then reinvested) in the City of London. Rubinstein (1977) has pointed out that between 1858 and 1914, out of 151 millionaires who died in the UK, 77 came from London and 49 of these hailed from the City. Similarly, out of 403 half millionaires, 167 came from London, and 117 of these came from the City (see also Rubinstein 1980).

In this final section, the argument is that much the same situation now pertains in the South East as during much of the nineteenth century and for reasons that are not dissimilar. That is, wealth and the demand for goods and services that this wealth represents still generate their own momentum in the South East, and amongst the chief beneficiaries are those working in the City of London.

Wealth is disproportionately gathered in the South East, whether in the form of earnings, total income, investment income, or accumulated assets. Thus, according to *New Society* (1986), 58 per cent of those earning over £50,000 in the UK lived in the South East, where the average earnings are generally much higher than in the rest of the country. Some 51 per cent of those with a total income of £20,000 in 1983/84 in the UK lived in the South East (Board of Inland Revenue 1987); 42 per cent of millionaires also lived there (*New Society* 1986).

Since 1979 wealth has increasingly been redistributed to the wealthy (see Rentoul 1987) and the number of moderately and very wealthy people has accordingly increased quite rapidly (see Thrift 1987b). For example, Shorrocks (cited in Rentoul 1987) calculates that the number of millionaires increased from 13,000 in 1983 to 20,000 in 1986. A number of reasons have been proffered for the increase in the ranks of the wealthy, including factors such as: rapidly increasing executive incomes; a favourable taxation environment; very favourable rates of return on capital; a bullish stock market (most people who are wealthy own a greater proportion of their assets in stocks and shares than the average for the population and so have benefited disproportionately from the rising stockmarket); low rates of inflation; and high house price inflation.

It would make a pleasingly symmetrical arrangement to argue that a spatial concentration of wealth in the South East has also occurred since 1979, over and above its already high levels of wealth, but there is only very limited data to confirm this argument. Certainly earnings have increased faster in the South East than elsewhere, especially for top earnings bands (Department of Employment 1986). The limited data on incomes (Board of Inland Revenue 1987) also suggest a degree of concentration. However, data on regional distribution of wealth are almost entirely lacking although one might assume that those with substantial assets have increased faster in the South East than in other parts of the country for various reasons, including: the greater preponderance of unlisted Stock Market firms in the South East than elsewhere; the presence of the City of London with its spiralling incomes and, of course, above-average house price inflation. Over the past twelve years, house prices in London and the South East have outstripped the national average by at least 25 per cent (Pitcher 1987). Set against this fact has to be the increased amount of income that is required to meet the generally higher level of mortgage repayments in the South East. Even so, members of households are accumulating tangible assets, which are backed by ready availability of mortgage loans.

The concentration of wealth in the South East has generated a buoyant economy compared with the rest of the country. It is an economy that is buoyed up by the consumption of goods and services. The average level of consumer spending in the South East is much higher than in other regions. Even allowing for a higher cost of living, its inhabitants spend more on almost every category of household expenditure. Proportionately more of their household expenditure goes on 'housing', 'durable household goods', 'other goods', and 'transport, vehicles, services and miscellaneous' than in most other regions. Similarly, much of the prosperity of the South East seems to come from services that are generated by income.

The City as a generator of wealth

What has been the role of the City in bolstering the dominance of the South East within the UK's spatial economy, especially in the lead up to, and after, the 'Big Bang'? It can be posited that, as in the nineteenth century, the City produced a disproportionate amount of wealth which then spread out through London and the South-Eastern economy, in the shape of demand for goods and services, and produced more jobs and more wealth in a 'virtuous circle'. This is not an easy argument to confirm without analysing considerable amounts of data and undertaking a series of local studies. Therefore what follows is a preliminary, impressionistic analysis which concentrates on personal wealth only. However, it is important to note the other impacts of the 'Big Bang' on the South East, which emanated from the corporate sector.

In 1986, the City's 390,444 workers had, because of their relatively high salaries, a gross income of about £5,423,980,000. Allowing for tax and other deductions, they were left with probably some £3,254,388,000 in actual income. (It should be noted that this figure is very conservative since it does not include indirect spending by firms in the City, on items such as company cars and private health schemes, or boosts to housing spending in the form of mortgage subsidy.) This money was spent all over the South East. Allowing for items such as reinvestment of income (often flowing back into the City), foreign travel (but booked through South-Eastern travel agents!), commuter travel, and expenditure in central London (on items such as consumer goods, the theatre, restaurants, clubs), we can make the assumption that the residue was spent at or near the home locations of professional and managerial workers, rather than clerical and other workers, and at the home locations of men rather than women, there being an obvious and unfortunate correlation still between professional/managerial workers and men, and clerical workers and women. A very crude idea of the spatial pattern of spending can be obtained by assigning average salaries of men and women in accordance with the spatial distribution of male and female workers in the City, which may be gleaned from 1981 census data. Not surprisingly, this brings the result that most disposable income is being spent in the better-off areas of London and the classical commuter belt settlements, so generating jobs, and more affluence, there. (This finding is confirmed by the information that it is possible to obtain on the home locations of professional and managerial City workers, in aggregate and by individual firms.) In turn, it might be expected that the affluence of these localities is partially connected with the spending power of City workers. (The exact relation will obviously vary according to the proportion and type of City worker to be found.)

On what was the income from the 'Big Bang' and its aftermath being spent? Obviously much was spent on essentials but the more wealthy City workers had considerable amounts of discretionary income to spend on goods and services. This discretionary income was being spent primarily on demonstrating 'taste' (Bourdieu 1984; Thrift 1987).

Spending on goods

Discretionary income tended to be spent either on status goods, such as cars, or on a subset of status goods, 'positional' goods (Hirsch 1977). These items, such as antiques and fine art, are scarce by reason of short supply and gain their value from being scarce. Commodities like these demonstrate the 'nobility' of their buyers.

The country house is a classical piece of positional goods. The market in country houses has boomed since the 'Big Bang' and without doubt much of the initial activity in this market could be traced to the new City rich, especially when the houses were in commuting distance of London:

> More than anything else, the 'Big Bang' in the City of London altered people's attitudes to country houses. For one thing it produced overnight a generation of young people who could afford to buy the houses that their elders had been struggling to keep up for years. For another, it magnified the importance of accessibility to London, resulting in the present distortion of property prices in the South East. The stockbroker who has to be in front of his SEAQ computer at 7 a.m. does not want to live too far away from London.
>
> (*Country Life* 1987:10)

The spatial distribution of the rise in country house prices since 1981 was mainly concentrated in the South East but it also spread along the M4 corridor (*Country Life* 1987).

A profile of typical country house owners, which was drawn in 1986, confirms expectations (Savills 1987). Most came from the City. They were relatively young, with most men in their 40s and most women in their 30s. They were well off, with 74 per cent earning more than £50,000 per annum and 41 per cent earning more than £100,000. Most were looking for houses in the £250-350,000 price range. Over half would purchase with cash. Those who had loans were often paying more than £2,000 per month for a mortgage.

Consumption on tasteful goods such as country houses demonstrates the 'cultural nobility' of the moderately and very wealthy. That taste is either upper class or middle class imitating upper class. Consumption is aimed at reproducing upper-class lifestyles through a constellation of goods, each of which reciprocally confirms the taste and background of the owner (Thrift 1987). Thus the country house is a shell for consumption of other goods: the Range Rover, the tweedy country clothes, the stable, and so on.

Spending on services

Not all discretionary income is spent on goods; much of it goes on services (Gershuny and Miles 1983). Of course, the consumption of tasteful goods can often imply the use of services, and there may be substantial multipliers even here. The purchase of a status car supports a network of distributors and repair shops. The purchase and upkeep of a country house might require an estate agent, an accountant, a construction company to do the renovations, plus cook, nanny, gardener, and so on. However, other services are also important. From limited research, much of the discretionary spending on services by more affluent City workers seems to be bent towards maintaining and reproducing upper-class (or middle class imitating upper-class) 'lifestyles' (see Ascherson 1986a; 1986b; 1986c). That might just mean appropriate restaurants but it can also mean education of children.

The case of private education is particularly instructive. Most affluent City workers send their children to a private school. Given the number of workers involved, children with City parents must have accounted for a considerable number of the 440,874 pupils being educated in private schools in the UK in 1986/7 (about seven per cent of the total UK school population). Some 34 per cent of these private school pupils were being educated in schools in Greater London and the South East. In turn, spending on private education keeps others employed; for example, in the UK as a whole, 33,248 teaching staff were employed in 1986/7 in private schools (Independent Schools Information Services 1987).

Consumption of services such as private education demonstrates how such services can be bent towards the reproduction of particular classes, in this case most prominently the upper and upper middle classes. The case of private schools is indeed exemplary. Routinely, private school pupils win about 50 per cent of Oxbridge places: in 1987, 46 per cent of Oxford places and 49 per cent of Cambridge places went to private school candidates (*Times Higher Education Supplement* 28 August 1987:1). Whilst at Oxbridge they are likely to be snapped up by City firms, a likelihood strengthened by their private school background. They are consequently paid high salaries, and so the round continues: class and region, region and class.

Conclusions: the region is alive and well, and living in the global system

In this chapter, I have tried to show that regional geography still has a place in the modern international world. The global systems of today can reproduce regions with distinctive economic, social, and cultural structures, as shown by the example of the 'Big Bang' in the City of London and its consequent effects on the South-East of England. Certainly the relationship between the global system and regions is now based on a more complex web of interconnections than ever before. But it might be argued, *pace* the arguments of a number of postmodern theorists (see Thrift 1989), that the very complexity of these interconnections, and the combinations of structure and action that result from them, means that greater heterogeneity is as likely to result as a tendency to homogeneity. Regional diversity has not disappeared, it has just taken on new forms which we are still struggling to unscramble.

Acknowledgements

I should like to thank Andrew Leyshon for his very considerable help with this paper.

References

Abercrombie. N. and Urry, J. (1983) *Capital, Labour and the Middle Classes*, London: Allen & Unwin.

Ascherson, N. (1986a) 'London's New Class: The Great Cash-In', *Observer*, 25 May.

— (1986b) 'The New Rich Spend, the Old Rich Sulk', *Observer*, 13 July.

— (1986c) 'Stoke Newington: Settlers and Natives', *Observer*, 3 August.

Ayling, D.E. (1986) *The Internationalisation of Stockmarkets*, Aldershot: Gower.

Banker (1968) 118:915-23.

— (1977) 127:129-77.

— (1986) 136:69-132.

Barrett, M. (1987) 'The Coming Crunch for Banks', *Euromoney*, April:99-103.

Board of Inland Revenue (1987) *Survey of Personal Incomes* 1983/84, London: Her Majesty's Stationery Office.

Bourdieu, P. (1984) *Distinction. A Social Critique of the Judgement of Taste*, London: Routledge & Kegan Paul.

Bowen, D. (1986) 'Class of 86', *Business*, November:34-41.

Browning, P.J (1986) 'Changes in the International Capital Market', in Zenoff, D.B. (ed.) *Corporate Finance in Multinational Companies*, London: Euromoney.

Census of Population 1981 Workplace and Transport to Work Tables (1984) London: Office of Population Censuses and Surveys.

Central Statistical Office (1987) *Regional Trends*, London: Her Majesty's Stationery Office.

Churchill, C. (1987) *Serious Money. A City Comedy*, London: Methuen.

Clarke, W.M. (1986) *How the City Works*, London: Waterlow.

Clifford, J.C. (1988) *The Predicament of Culture. Twentieth Century Ethnography, Literature and Art*, Cambridge, Mass.: Harvard University Press.

Cobbett D. (1986) *Tales of the Old Stock Exchange. Before the Big Bang*, Portsmouth: Milestone Publications.

Country Life (1987) *Buying a Country House. A Regional Guide to Value*, London: Country Life.

Department of Employment (1986) *New Earnings Survey*, London: Her Majesty's Stationery Office.

Economist (1987a) 'Nice Market, Shame about the Backlog', 29 August:71.

— (1987b) 'Big or Boutique?', 29 August:16-17.

Employment Gazette (1987) (January), Department of Employment, London: Her Majesty's Stationery Office.

Euromoney (1987a) 'The Most Powerful Men in Japan', March: 118-23.

— (1987b) 'The Power League', February:85-95.

— (1987c) 'Swaps: New Moves', Supplement, July.

Eykyn, G. (1987) 'Biz Gets the Grads', *Observer*, 24 May 1987:49.

Family Expenditure Survey (1985) London: Her Majesty's Stationery Office.

Farmborough, (1987) 'Atomic Reaction', *Financial Weekly*, 18 June:28-34.

Findlay, A. (1987) 'From Settlers to Skilled Transients. The Changing Nature of British International Migration', Paper presented to the Workshop of the Skilled International Migration Working Party, University of Liverpool, 6-7 July 1987.

French, M. (1987) 'Swapping Debt: Just Hot Air', *Euromoney*, May:115-22.

Gershuny, J. and Miles, I. (1983) *The New Services Economy*, London: Frances Pinter.

Goodhart, C.A.E. (1987) 'The Economics of Big Bang', *Midland Bank Review*, Summer:6-15.

Hamilton, A. (1986) *The Financial Revolution. The Big Bang Worldwide*, Harmondsworth: Penguin Books.

Hamilton, J.D. (1986) *Stockbroking Tomorrow*, London: Macmillan.

Hammond, G.S. (1987) 'Recent Developments in the Swap Market', *Bank of England Quarterly Bulletin*, February:66-74.

Hewlett, N. and Toporowski, J. (1985) *All Change in the City*, Special Report No. 222, London: Economist Publications.

Holberton, S. (1987) 'Tokyo on the Thames', *Financial Times*, 25 July.

Hirsch, F. (1977) *The Social Limits of Growth*, London: Routledge & Kegan Paul.

Independent Schools Information Service (1987) *Annual Census 1987*, London: Independent School Information Service.

Ingham, G. (1984) *Capitalism Divided*, London: Macmillan.

Ingram, D.H.A. (1987) 'Changes in the Stock Exchange and Regulation in the City', *Bank of England Quarterly Bulletin*, February:54-65.

Kaufman, H. (1986) *Interest Rates, the Markets and the New Financial World*, London: I.B. Tauris.

Key, J.S.T. (1985) 'Services in the UK Economy', *Bank of England Quarterly Bulletin*, 25:404-14.

Labour Research (1986) 'Divided City', November:11-14.

Lash, S., Urry, J. (1987) *The End of Organised Capitalism*, Polity, Cambridge.

Lawless, J. (1987) 'Back from the Brink', *Business*, May:84-6.

Lee, C.H. (1984) 'The Service Sector, Regional Specialisation and Economic Growth in the Victorian Economy', *Journal of Historical Geography* 10 (2):139-55.

Leyshon, A., Thrift, N.J., and Daniels, P.W. (1987) 'The Urban and Regional Consequences of the Restructuring of World Financial Markets: The Case of the City of London', *Working Papers on Producer Services*, No. 4, University of Bristol and University of Liverpool.

Lim, W.P. (1986) 'The Euromoney Risk Ratings', *Euromoney*, September: 364-9.

Marsh, F. (1986) *Japan's Next Export Success: The Financial Services Industry*, Special Report No. 1066, London: Economist Intelligence Unit.

May, M. (1987) 'City Reward that Equals £111,000', *The Times*, 21 July:25.

New Society (1986) 'The Rich in Britain', *New Society*, 22 August, Special Supplement.

Nomura Research Institute (1986) *The World Economy and Financial Markets in 1995*, Tokyo: Nomura Research Institute.,

OECD (1986) *Financial Market Trends*, November.

O'Leary, J. (1986) 'The Powerhouse Beckons with a Pot of Gold', *Times Higher Education Supplement*, 20 June:12.

Pagano, M. (1986) 'Graduate Bank on the Big Bang Bucks', *Guardian*, 12 August:22.

Plender, J. (1987) 'Under the Skin of an Image Problem', *Financial Times*, June:15.

— and Wallace, P. (1985) *The Square Mile*, London: Century.

Price, K.A. (1986) *The Global Financial Village*, London: Banking World.

Regional Trends (1983) London: Her Majesty's Stationery Office.

— (1986) London: Her Majesty's Stationery Office.

Remuneration Economics (1987) *Guide to Current Salaries in Banking*, March, London: Jonathan Wren Recruitment Consultants.

Rentoul, J. (1987) *The Rich Get Richer. The Growth of Inequality in Britain in the 1980s*, London: Unwin Hyman.

Rubinstein, W.D. (1977) 'The Victorian Middle Classes: Wealth, Occupation, and Geography', *Economic History Review* 30:602-23.

— (1980) *Men of Property*, London: Croom Helm.

Savills (1987) *Savills Magazine* 16, Spring.

Taylor, A. (1987) 'Salaries Low, Office Costs High in London', *Financial Times*, 13 May.

Thrift, N.J. (1986) 'The Internationalisation of Producer Services and the Integration of the Pacific Basin Property Market', in M.J. Taylor and N.J. Thrift (eds) *Multinationals and the Restructuring of the World Economy*, London: Croom Helm.

— (1987a) 'Manufacturing rural geography?' *Journal of Rural Studies*, 3:77-81.

— (1987b) 'Serious Money. Capitalism, Class, Consumption and Culture in Late Twentieth Century Britain', Paper presented to the IBG Conference on 'New Directions in Cultural Geography', London, September.

— (1987c) 'The Fixers: The Urban Geography of International Commercial Capital', in J. Henderson and M. Castells (eds) *Global Restructuring and Local Areas*, Beverly Hills: Sage.

— (1989) 'Taking Aim at the Heart of the Region', in D. Gregory, R.L. Martin and G. Smith (eds) *Geography in the Social Sciences*, London: Macmillan.

— and Leyshon, A. (1988) 'The Gambling Propensity', *Geoforum*, special issue on the debt crisis 19:55-70.

Townsend, P. (1987) *Poverty and Labour in London*, London: Low Pay Unit.

Epilogue: towards an agenda for regional geographical research

Joost Hauer, Gerard A. Hoekveld, and R.J. Johnston

About basic elements in regional geography

Marc de Smidt, in his concluding comments at the end of the symposium in Utrecht, stated that the different philosophical and ideological viewpoints that are presently current in geography affect the approaches of regional geographers. Thus they lead to different outcomes of regional geographical analysis and description. De Smidt noticed in some papers, as well as in the ensuing discussions, a clear tendency to reduce the contrasts between Marxian, institutional, behavioural, neoclassical, humanistic, and other views to the contrasts between a more interpretatively directed and a more structuralistic view. About the last he said:

> The bridge built by Anthony Giddens in his structuration theory in interrelating agents and systems via institutional arrangements (including the planning arena) does not imply that the different theoretical views have to be pulled back. The function of a bridge can be interpreted as representing a framework of discussion by social scientists of different backgrounds instead of splitting up debates on realities along lines of so-called schools of thought.

De Smidt stressed the primordial importance of scale in regional geographical research, particularly in research within the structurationist mould: 'The basic problem of regional geography is how to select the interdependent elements on the same as well as on different scales. This is a specific craft the regional geographer has to master'. He distinguished at least three important, interlinked scales among the many possible. There are those of the so-called 'agents' spaces' or 'habitats', the 'territories', and the 'systems'! The agents' spaces are based on the fundamental notions of coexistence and mobility that underlie the analysis of time-space trajectories of 'actors' or 'agents in space'. These trajectories bring about different habitats: bounded spaces which are regularly used by individuals and groups. The segmentation of society in many different categories of population and the nationalization or internationalization of the economy differentiate space on a local scale in ways that have direct impacts on the daily existence of individuals and categories of population.

The scales of the territories that are the bounded spaces in which organizations perform their functions are also very important. 'Territoriality is the geographical expression of the intermediary role of institutions between agents and systems. ... Institutions are intermediaries by the ways they structure time-space trajectories of people belonging to the different social categories that make up regional or local populations.'

Systems, conceived by De Smidt as interregional and international, need not necessarily be geographical phenomena. They require, however, regional assets for their locations and interactions as a base for their operating on a supra-regional scale. Their world-wide factor markets and markets for goods, as well as their regional bases, e.g. the world cities, are interlinked with the agents' spaces and territories of other institutions because they share the same areas or parts of them at some place. Therefore a regional perspective in human geography has to deal with all three basic elements of analysis, agents, institutions, and systems, at the same time and, consequently, with their accompanying societal and spatial contexts. In a flexibly devised structuralistic framework, that condition might be satisfied and room might be left for interpretative analysis of the agents' activities. This brings us to the question of what a new regional geography might look like.

Some characteristics of developing regional geography

It has been observed that there is no 'natural' discipline of geography; instead, geography is continually being recreated by the practices of those who call themselves geographers (Johnston 1985:25). This is particularly true for regional geography. The history of this discipline shows that it follows the same trends in time and space as the other fields of geography through the 1950s. Since then, however, its development has stagnated. Whereas the topical geographies have been renewing their theoretical, methodological, technical, and substantive bases, regional geography seems to have come to a standstill in that respect.

Regional geography is being revived, but it should be more than a fad, more than the revival of a glorious past. It should bring older geographers, with their experience, into contact with younger geographers who will work to ensure that this new regional geography will give content and relevance to the discipline today. Reviewing the consequences of stagnation can be a major input in making regional geography a relevant and concrete part of the discipline.

In this book, several contributors have pointed out some causes of this stagnation and there is no need to repeat them here. However, it may be useful to explore the consequences of the stagnation for the subject matter and for the form of a new regional geography. Of course, these consequences differ according to the directions that are indicated in the contributions to this book but there seems to be agreement among the authors on some characteristics of the traditional discipline which are slated for renewal and which distinguish it from the 'new' regional geography that was sketched in the introduction.

The first of these characteristics is the low profile of traditional regional geography among the subdisciplines. The demise of regional geography has undoubtedly been hastened by the high ideals that are held by many practitioners of traditional regional geography. It was supposed to be the culmination of geographical endeavour, the great synthesis of geographical thought.

A main characteristic of the new regional geography is its rejection of the claim to study individual regions that cannot be accounted for by general principles or generalizations. The search for the identity or personality of regions has directed too much attention to singularity. Leading classical geographers, such as Vidal de la Blache and Hettner, always emphasized that regions were not singular but unique (i.e. a unique combination of general principles of a given place and time).

The generalizations that are sought and investigated are not natural laws but historical abstractions of societal development. This is connected with the anthropocentric approach to the region that is peculiar to regional geography. Classical geography and regional geography assumed that the regions were an intricate product of the combination of natural and human forces. At present, space is considered to be a realization of human perceptions, ideas, intentions, and technology. Consequently, people may disengage themselves from local or regional constraints by their capacity to organize localities as parts of larger wholes. History shows a continuous evolution of this capacity. Regional geography is therefore compelled to come to grips with the development of the larger societies that are comprised of regional populations.

Historical development is not the '*histoire événementielle*', the history of the succession of events in a society. Rather, it is the history of processes underlying the events in which history manifests itself. These processes are the focus of those who try to establish generalizations of societal development and human action within historically and spatially limited ranges of validity. This implies a new way of handling history. History previously served as a genealogy of the successive transformations, additions, and eliminations of the region's traits and structures. Now it serves as a backdrop for general developments.

The fourth characteristic, concreteness, may not be so new, but it is very important. Regional geography will have to make a commitment to concreteness if it is to fulfil its social and scientific aims. This has implications for the concept of the region. The region is not the abstract space that is produced by many students of topical geography but the unique place in which historical processes become the spatio-temporal realities of daily life. 'Man' in regional geography is no longer hidden behind 'the group' but may be an individual, perhaps aggregated in categories, perhaps as part of groups or organizations, but not necessarily so. This concern for concrete people allows regional geography to look for abstract structures and processes behind or in the individuals and their places, but does not allow regional geography to limit itself to that.

A fifth difference between old and new regional geography lies in its use of a broad array of methods and techniques for description and analysis. The old, yet indispensable, techniques of visual inspection, description, and surveying are complemented by many kinds of description and analysis which are adopted from the social sciences. The techniques allow an investigator to penetrate the subject at all levels of analysis. At the moment, there are no well-established guidelines for regional description and analysis, although these did exist for many years before the Second World War in most European countries. However, there is a growing recognition that description and analysis should be done at the level of individuals as well as that of organizations or groups and aggregates. Traditional simple or sophisticated statistical techniques may be supplemented by participant observation, hermeneutical-content analysis, and other similar techniques. This expansion of the technical base of regional geography does not imply that regional geography 'went technical' to prove its scientific status. On the contrary, a wider range of techniques is necessary to enable regional geography to remain, or to become again, a geography of the concrete.

A lively debate in Germany was started when Birkenhauer (1970) published his paper, 'Regional Geography is dead, long live regional geography', in reply to Schultz's paper, 'General geography instead of regional geography'. It was argued that many young geographers had declared that regional geography was as dead as a doornail, but that there was a need for regional knowledge. Geography should provide for these wants by finding new ways of studying regions. This probably characterizes the situation in Germany in the 1970s, but it is not enough now to confess that we dropped a stitch in our enthusiasm to become real scientists, nor to claim that society needs a good regional geography in order to solve its areal problems and its global interdependencies.

Regional geography has missed many good opportunities. Although geographical monographs continued to be published, other sciences hastened to fill the void that was left by the retreating geographers. Regional history widened its scope; no longer limiting itself to the past, it often included the contemporary situation. For lack of adequate access to primitive people in non-Western countries, cultural anthropologists found niches in the regions of the Western world. Many journalists published on subjects that had previously been the domain of regional geographers. Of course, a vital regional geography can reclaim that lost territory but regional geography should do more. It should work on new research frontiers which have emerged with the changed state of affairs. New research directions have also been indicated by the geographical subdisciplines as a consequence of their differential progress during the past three or four decades.

Research Frontiers

What are the principal research frontiers for the immediate future? One is the methodology of regional geography. Regional geography informs, it sketches portraits of regions. Every publication, film, or other form of documentation of a region is necessarily incomplete; it reduces, magnifies, simplifies, and makes choices among factors which are considered to be more (or less) important. This is a normal procedure in scientific work but, in a concrete discipline such as regional geography, it is a very delicate process. Even the public for whom a regional geography is written may influence the choices that an author makes in anticipating the interests and knowledge of the readership. Methodological reflection, which includes assessing the consequences of the ideological stances of the authors, is crucial to the resulting portraits of the regions and to the way in which regional characteristics are ascribed to processes taking place in regions. This research frontier may be described as one of methodology. How should different levels of analysis be linked? Which transformations of classes (types or concepts) must be executed? These answers are preliminary to the study of regions and subregions that purports to disclose both the general and the unique traits of regions at one and the same time.

We still have not dealt with the fact that geography studies populations and societies and, as such, is a social science, employing the theories and methodologies of the social sciences. At the same time, geography relates these populations and societies to a concrete area and a specific spatial content. We have not given enough thought to the consequences of this fact for our theories and our methodological choices.

A second research frontier has a very concrete nature. Modern regional geography is context-bound. 'External' relations between regions are often seen as more important than 'internal' relations, that is, relations between components of the regional system. The largest context is the world-system, or the new international division of labour. At the moment, we are fully aware of the importance of the world-system. It is the fabric of external relations that elicits regional responses or adaptations in the form of spatial developments. We do not know much about the spatial structures of this world-system, however. 'The' system may consist of spatial subsystems of economic and political subsystems, but we do not understand how these subsystems relate to each other nor how they 'work'. Geographers have not yet learned to study the accumulation and use of capital in a geographically appropriate way. Nor have we also yet discovered how the world-system, the nation states, the world cities, and the other regions interact, although we know that the multinational enterprise plays a key role. Historians such as Braudel and sociologists such as Wallerstein have paved the way but the geographers seem to lag behind.

The third research frontier is the problematic one of regional development. Of course, an enormous amount of work has already been done by distinguished scholars such as Friedmann, Weaver, Perroux, Massey, and Fischer but there is still more work to be done in regional geography. Some

Marxist geographers and many regional economists, who advocate development 'from below', have arrived at the insight that socio-cultural structures and processes are crucial to regional economic development. Others substitute the concepts of urbanization, core, and periphery for complex regional development in an attempt to avoid economic reductionism. Until now, regional development theories have not been sufficiently connected with spatially differentiated and contextually disparate regions. Partly this may be a consequence of the theories and their analytic procedures being too abstract to comprehend the complex and varied regional reality. Partly this has to do with a problem that Scott and Storper have raised (1986:10). They think that an understanding of territorial complexes such as regions requires a total reassessment of the theories of the social sciences:

> A major puzzle for all the social sciences is therefore how historical eventuation — that is, the historical development of capitalism as a whole — is played out through the intervening effects of locale. We suggest that the analysis of modern capitalism in terms something like these would significantly augment the whole corpus of contemporary social theory. A more forceful way of making the same point is to suggest that the largely spaceless social theories that currently dominate our intellectual culture stand in urgent need of retotalization via the dimensionality of geographical space and the particularities of space.
>
> (Scott and Storper 1986:310)

This venture will probably be the theoretical framework into which the other developments have to fit. Regional development studies should be carried out in a wide range of regions, e.g. urban core, peri-urban, peripheral, and border areas, if possible in a comparative way. Border areas in particular may suggest interesting generalizations about regional development in the context of the nation-states and the world-system. They teach us how external relations can be mitigated or stimulated, and how centre-periphery relations operate. This work often allows interesting comparisons of regions in different countries.

A fourth frontier was particularly stressed by De Smidt. It is the study of regions as meaningfully bounded spaces for different categories of the population into which modern, segmented society is divided. Historical developments may even have quite different impacts on the same categories or groups in different regions and thus may contribute to cultural, institutional, and political varying regional characteristics. Different compositions of social categories, too, may result in such divergent characteristics. It is important to know which traits of the environment, institutions, and modes of production maintain or alter these social regional qualities because they can explain the regionally unlike responses to exogenous efficacities.

The fifth frontier is the cultural identity of regions. Of course much attention has been focused on regions with separatist or devolution movements, such as Brittany, Scotland, Corsica, and South Tyrol. The problems of cultural identity are more disguised in many other regions, but they are there none the less. The cultural identity of regions in an urbanized world is articulated by groups and organizations; it is supported by the media, by special interests, by politicians, etc. The problem of regional identity is related to the adaptation of regions to the world-system. Their populations should be conscious of their identity and their aspirations. Regional identity is related to emancipatory aims. Identity exists at different scales. The debates about the existence of Central Europe and about European education (or education for Europeans?) are no less important than the confrontation between regional organizations and local administrations acting on behalf of local communities. Geographers seem to have lost the capacity to depict a region as a cultural entity. Others, however, are still publishing such studies, for instance Garreau (1982) on North America, and Allemann (1985) who provides an insightful portrait of the Swiss cantons. Allemann analyses the historical evolution of these small provinces, specifically with respect to their contemporary socio-cultural, economic, and political characteristics. He sketches a basically static picture. Garreau, on the other hand, who treats the regionalization of North America, focuses on new fluctuating interaction patterns that are changing the socio-cultural and economic characteristics of the regions. Both studies, each from its own point of view, demonstrate the connectivity of regional development, regional differentiation, external relations, and culture. The geographer has difficulty in dealing with culture: recognizing its importance but lacking the tools to describe it adequately (i.e. avoiding stereotypes, too much detail, and too much eclecticism).

The sixth research frontier is not a direction but a research problem. We badly need models of regional differentiation. There are many models of spatial differentiation of cities. These include those of Schlüter (Germany), Burgess, Hoyt and Ullmann (USA), and the recent models of Lichtenberger

(Austria), Sabelberg (Germany), and Ashworth (The Netherlands), to name some well-known authors. Regional
geographers may lack models because they are wary of
nineteenth-century physical determinism and because they do
not adequately evaluate the natural conditions of the land.
Their hesitation can be overcome by bringing back an
anthropocentric physical geography into regional geography.
On that foundation we can develop typologies of 'natural' and
'morphological' areas that permit a confrontation with
historical stages of societal development, one such sequence
being feudalism, mercantile, industrial, and modern capitalism.
Irrespective of the typology used, each successive stage creates
its own spatial centres and peripheries. Moreover, each stage
corresponds to a type of spatial structure which is engendered
by mechanisms specific to the era. These morphological and
historical typologies of societal development may provide a
basis for generalizations on regional development and regional
structures that apply in the context of the uniqueness of each
region. These typologies may form a basis for models of
spatial differentiation that provide a frame of reference for
the study of individual regions. The development of these
models will be dependent on the progress that is made at other
research frontiers.

References

Allemann, F.R. (1985) 26 mal *die Schweiz*, Zurich: Piper.
Birkenhauer, J. (1970) 'Die Länderkunde ist tot; es lebe die Länderkunde',
    Geographische Rundschau 22:197-203.
Garreau, J. (1982) *The Nine Nations of North America*, New York: Aron.
Johnston, R.J. (1985) 'Introduction: exploring the future of geography',
    pp.3-26 in R.J. Johnston (ed.) *The Future of Geography*, London:
    Methuen.
Schultre, A. (1970) 'Allgemeine Geographiez statt Landerkunde', in
    Geographische Rundschau 22:1-10.
Scott, A.J. and Storper, M. (eds) *Production, Work, Territory: the
    Geographical Anatomy of Industrial Capitalism*, Boston: Allen & Unwin.